Monte Carlo Simulation in Dependability Analysis

Reliability of Multiphysical Systems Set

coordinated by
Abdelkhalak El Hami

Volume 21

Monte Carlo Simulation in Dependability Analysis

Franck Bayle
Laurent Denis
Adrien Gigliati

iSTE

WILEY

First published 2025 in Great Britain and the United States by ISTE Ltd and John Wiley & Sons, Inc.

ISTE Ltd
27-37 St George's Road
London SW19 4EU
UK

www.iste.co.uk

John Wiley & Sons, Inc.
111 River Street
Hoboken, NJ 07030
USA

www.wiley.com

Library of Congress Control Number: 2025944643

British Library Cataloguing-in-Publication Data
A CIP record for this book is available from the British Library
ISBN 978-1-83669-032-0

The manufacturer's authorized representative according to the EU General Product Safety Regulation is Wiley-VCH GmbH, Boschstr. 12, 69469 Weinheim, Germany, e-mail: Product_Safety@wiley.com.

Contents

Foreword by Philippe Bogdanik

There are first meetings that leave a lasting impression. My first meeting with Franck Bayle at the end of the 20th century struck me at the time. I was a student and had just arrived for an internship at a company, where we would work side by side for many years until his retirement. It was also my first meeting with my internship supervisor, also the head of department, who brought us together with Franck to explain the activities and studies to be carried out to develop new power electronics intended for the aeronautical world. A monologue ensued from the electronics expert, who gave us his instructions and then dismissed us to let us organize our work. Then, Franck asked me seriously, "Did you understand what he said? Can you explain it to me?" I admit I was surprised by his question: how could an engineer, expert his field, in a large group not understand what his superior was saying?

After several decades, this questioning seems natural coming from Franck. He obviously understood, but only a few subtleties had escaped him. He has always been curious to understand the world around him. Now it is up to him to enlighten me with his knowledge.

After abandoning electronic development to focus on reliability studies, he has always sought to improve his knowledge and then develop working methods, which has not been easy in the company. The proposed change, if it brings a certain additional cost compared to future gains which cannot be immediately determined, is difficult to adopt, especially if the additional costs occur during the development phase and the gains during the production or maintenance phase.

Paradoxically, Franck was more recognized outside the company, where he was able to forge links with other experts in the reliability field. He always sought to confront the best in his field, which makes him an unusual case among experts.

Now, it is time for Franck to complete his transmission of knowledge accumulated over the years by bringing in his wake Laurent Denis and Adrien Gigliati, who he has worked with for years. The numerical resolution methods they have developed will find, through this book, an audience informed about reliability issues.

Philippe BOGDANIK
Engineer at Thales AVS France
September 2025

Foreword by Gilles Zwingelstein

Operational safety (OS) is a major concern for those responsible for operating complex industrial systems to meet operational and regulatory requirements. OS is a multidisciplinary scientific discipline in its own right and is constantly evolving to integrate new tools and requirements, particularly in the areas of maintenance and technological risk management. It is defined as "the ability of an entity to satisfy one or more required functions under given conditions". In a broad sense, OS is considered the science of failures and breakdowns. The major difficulties in operational safety lie in the random nature of the moment of occurrence of the failure, the knowledge of its reliability law and the difficulty of carrying out real reliability tests. Many explicit or statistical methods have been developed by many researchers.

This book exhaustively describes the Monte Carlo simulation methods in operational safety. It clearly describes, for the non-specialist, the reasons why, Monte Carlo simulations are essential when analytical methods are insufficient or impractical. It presents four parts that are essential to technicians and engineers concerned with Monte Carlo simulation methods for reliability studies, spare stock management, availability and safety. This book differentiates the cases of systems operated without maintenance from those subject to maintenance. It proposes a very relevant classification of reliability during the lifetime of equipment by describing the problems linked to predictive reliability and operational reliability by describing the role of Monte Carlo simulations.

This book differentiates the case where explicit solutions exist from that where no analytical solution is possible. Also, a very relevant and detailed analysis is provided on reliability testing. Indeed, this area is the Achilles heel of dependability studies because it is very difficult to experimentally validate failure rates of components with FITs of billions of operating hours. The book proposes carrying out a "virtual" reliability test by Monte Carlo simulations to try to solve this crucial problem for high-tech industries. The end of the book presents the applications of Monte Carlo simulation methods for the availability and safety of systems in the presence of failure.

In conclusion, this book is aimed at engineers, researchers and industry professionals seeking to deepen their knowledge of system reliability and Monte Carlo simulations. This book provides clear answers and practical tools to address the challenges posed by the complexity and uncertainty inherent in modern technological systems.

Gilles ZWINGELSTEIN
University Professor
September 2025

List of Notations

AF	Acceleration factor
D	Random variable modeling repair times
E[X]	Mathematical expectation of the random variable X
Ea	Activation energy
T	Random variable modeling times of correct operation (uptimes)
f(t)	Probability density
F(t)	Failure probability
h(t)	Instantaneous failure rate
H(t)	Cumulative failure rate
Mod	Modulo
MTTF	Mean time to failure
N(t)	Number of failures observed at time "t"
R(t)	Reliability function
Rocof	Rate of occurrence of failure

$Rocof_{\infty}$	Rocof asymptotic value
β	Shape parameter of the Weibull distribution
η	Scale parameter of the Weibull distribution
θ	Temperature
λ	Parameter of the exponential distribution
λ(t)	Failure intensity
φ	Probability density of the normal distribution
ϕ	Probability of failure of the normal distribution
μ	Scale parameter of the Weibull distribution
σ	Shape parameter of the normal distribution
ב	Indicator function

List of Acronyms

AFT	Acceleration failure time
ASIC	Application-specific integrated circuit
FIT	Failure in time (10^{-9} failure/h)
FMEA	Failure mode and effect analysis
FTA	Fault tree analysis
HPP	Homogeneous Poisson process
MTBF	Mean time between failure
MTTF	Mean time to failure
MTTR	Mean time to repair
PLP	Power law process
REX	Return of experience
ROCOF	Rate of occurrence of failure
TTF	Time to failure
TTR	Time to repair

Definitions

Alternating renewal: renewal process where repair times are not zero.

Element: sub-assemblies, electronic boards or assembled components in a system.

Equipment: assembly of interacting sub-assemblies.

Ideal maintenance: maintenance that instantly renews an element or system.

Renewal process: the function of a renewal process is to count the occurrences of a given phenomenon, when the time between two consecutive occurrences is independent and identically distributed random variables. This may involve counting the number of breakdowns of electronic equipment in the theory of reliability (the equipment is then renewed after each failure, hence the name), counting the arrivals of customers in a queue, survey the instances of a disaster for an insurance company, etc.

Simple renewal: renewal process where repair times are zero.

System: a group of elements capable of performing or supporting an operational role. A full system includes all of the equipment, hardware, software, services and personnel necessary for its operation so that it can be self-sufficient in its operating environment.

Introduction

When Napoleon asked Laplace why his treatise on cosmology did not mention God, the latter replied, “Sire, I had no need for this hypothesis”. At the end of the 19th century, physicists thought they had covered the whole of physics, and it was considered purely deterministic. There was however a small problem, but they thought it would be quickly resolved during the following century. Indeed, Young’s famous slit experiment has remained somewhat forgotten. The problem encountered was that, depending on the conditions of the experiment, light behaves either like a particle or a wave, and there was no explanation for this at the time.

Owing to Planck, Einstein, Bohr, Heisenberg, Pauli, Dirac, Schrödinger and Born, quantum mechanics was born in the 1920s. It essentially states that at the subatomic level, we cannot know the position of a particle with great precision. Rather, only its probability of presence and determinism give way to uncertainty. Despite the fact it is little addressed in educational programs, quantum mechanics is nevertheless present in our daily lives. It fuels the doubt that makes us uncomfortable, and we all seek to reduce, or even better, eliminate it. We continued to monitor weather forecasts, depsite the fact we know that the weather is notoriously unpredictable and that forecasts are quite often inaccurate.

In industry, we more commonly speak of operational safety. This discipline, which acquired this name and its current form mainly over the last half century and in the defense, aeronautics, space, nuclear, telecommunications and transport sectors, is now useful, even essential, for all sectors of industry and even other activities. The goal that requires the use of operational safety is more recognizable under the term “risk control”. Operational safety is, according to Villemeur (1988), “the ability of an entity to satisfy one or more required functions under given conditions”. It mainly encompasses four components: *reliability*, *maintainability*, *availability* and *safety*.

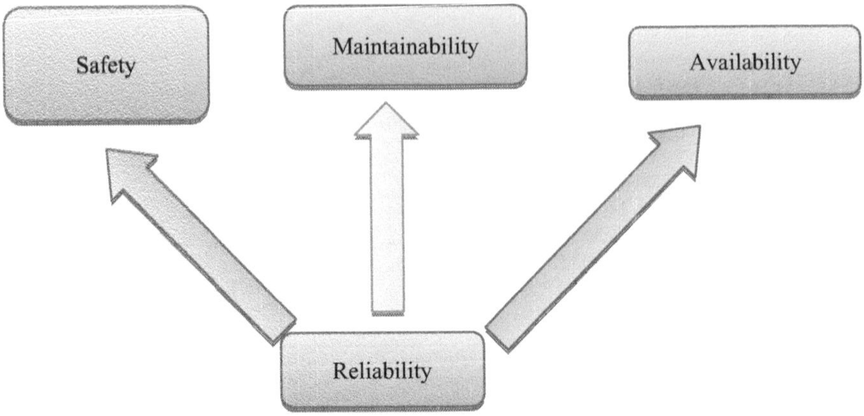

Figure I.1. *Various analyses of operational safety. For a color version of this figure, see www.iste.co.uk/bayle/montecarlo.zip*

Reliability is thus the quantitative basis for the other three analyses of operational safety. It can itself be divided into three distinct levels depending on the phases of the system's lifecycle. This can be illustrated synthetically by Figure I.2.

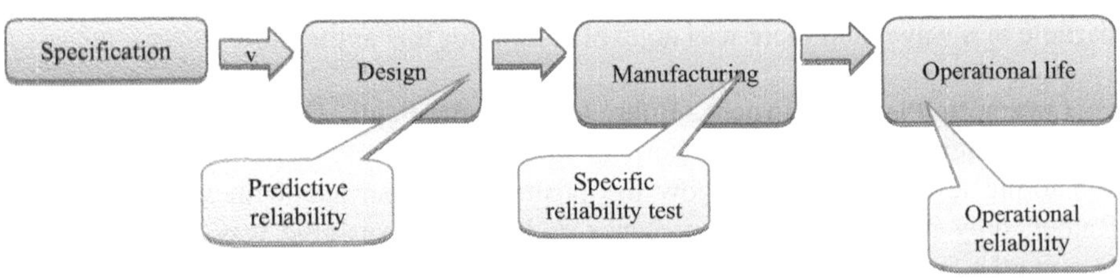

Figure I.2. *Different types of reliability and their positioning in the lifecycle of a system. For a color version of this figure, see www.iste.co.uk/bayle/montecarlo.zip*

The *maintainability* of a system is also a major parameter. Of course, it only makes sense for systems subject to maintenance. It does not depend directly on reliability (the reason for the green arrow in Figure I.1). For a system to be optimally maintained, it is first essential that the spare stock is in line with the number of system failures. Thus, the estimation of the "right" number of systems in stock can be assessed on the basis of the reliability level.

Availability is a critical parameter for many industrial applications, such as aeronautics, railways, energy production and distribution. Availability represents the ability of a system to perform a required function under given conditions at a given time. It depends not only on the reliability of the system but also on its ability to be repaired. The number of repairs is generally random.

Safety, more commonly referred to as "*safety*" in certain industrial fields, is the ability of a product to respect, during all phases of a device's life, an acceptable level of accident risk likely to cause an attack on personnel or major degradation of the product or its environment. It is generally broken down into an analysis of failure modes and their effects and an estimation of the probability of occurrence of certain feared events specified by the client.

The authors, despite sometimes working in different industrial fields, have very often encountered methodological errors, resulting in the confusion of the major parameters of these different areas of operational safety. For example, the failure rate is often referred to as a key reliability indicator, particularly on the Internet, whereas its practical use is restricted to a few industrial applications, such as space or certain military applications (missiles). It is also possible that, for a new generation component, reliability tests are carried out on test benches to verify that its intrinsic reliability is indeed as expected. However, these tests are rare because they are time consuming and have a significant cost. The effect of maintenance, which is present in most industrial applications, is often neglected even though it plays a fundamental role in the reliability of an operating system.

There are books detailing in a very rigorous way the theoretical methods used to approach these different themes, but they often remain very academic and difficult to master in the industrial field. In addition, we often address theoretical equations that do not have explicit solutions. In some cases, numerical solutions exist, but they can present certain problems, such as inconsistent results, without really realizing them, and they do not allow us to directly understand which parameters intervene in the variable of interest.

However, some renowned scientists often discovered revolutionary theories simply through thought experiments, well before these theories were experimentally verified. Therefore, it seems quite legitimate to want to simulate random events such as moments of failure, repairs or, more generally, the appearance of the feared or hoped-for fact under the assumption of probability distributions. *Monte Carlo simulations* can be an effective tool for resolving the various problems mentioned above.

The objective of this book, in an industrial context of operational safety, is therefore twofold:

– We propose theoretical solutions adapted to the problems encountered.

– The corresponding Monte Carlo simulations are presented when explicit solutions are not available or to verify the proposed theoretical approximations or numerical calculations.

In accordance with Figure I.1, the structure of the book is composed of four distinct parts. For each of them, we present a theoretical analysis for systems operated without maintenance as well as a theoretical analysis and Monte Carlo simulations for systems with maintenance. If an explicit solution exists, we use only this one. If, on the other hand, only a numerical solution is possible, we will then use simulations to corroborate the results obtained. Finally, if no analytical solution is possible, either because it does not exist or because it is not known to us, we will systematically use simulations either by the failure probability inversion method or by the rejection method when the reciprocal function has no explicit expression. The simulations carried out are either created in Python code or using the "Weibull++" or "BlockSim" software from ReliaSoft®.

We begin with Part 1 on *Reliability* because if there is one area where uncertainty is very present, it is that of reliability. Regardless of how well we know the physics of the technology used, the failure times of technological entities being tested or operated are always random.

During the so-called specification phase, a reliability objective is generally required by the client, which translates either into a probability of successfully completing a specific mission for systems operating without maintenance or into an "MTBF" for those with maintenance. Thus, at the end of the design phase, since the final (series) version of the system is not yet available (only prototypes are functional), a predictive reliability analysis is carried out to verify that the objective is met. The realism of this predictive analysis is therefore very important, particularly the estimation of the levels of physical contributions (life profile) to which the system will be exposed during its operational life.

Thus, Chapter 1 propose to evaluate the sensitivity of the forecast reliability estimate to the parameters of the life profile of a system. Therefore, instead of constructing it from constant levels, we propose, when necessary, to assume a probability law chosen according to the information collected. We then use Monte Carlo simulations to estimate the variability of the system reliability.

On the other hand, in certain specific cases, reliability tests on a component deemed "at risk" (with respect to the reliability of the entire system) may be deemed necessary or even essential. This may be a component with an insufficient reliability level on a previous generation and that has undergone a design change, a new technology component, a single-source component for which the manufacturer does not provide reliability information, etc. In contrast to the general case where we want to minimize the number of system failures, particularly during the design phase, we will therefore do everything possible to induce failures to obtain an "intrinsic" reliability model of the component with a given accuracy. The notions of bias and variance of the estimators of the reliability model parameters are then of the utmost importance. Chapter 2 therefore proposes a theoretical approach with explicit solutions when possible and Monte Carlo simulations when this is not the case.

Finally, so-called "operational" reliability (also called feedback or "REX") is that of the systems when they are in operation at end users. For systems without maintenance, this analysis is identical to that of a reliability test since there are no maintenance acts. In the majority of cases, maintenance acts are carried out during system failures. When a component fails, if the system is considered a "serial system", this leads to the failure of the system. The component, which is not a repairable entity, is then replaced by a new one. Thus, if the system contains a fairly large number of components (which is generally the case), we can make the assumption of so-called "minimal" maintenance since we have simply returned the system to the reliability level it had just before the failure. To model this type of maintenance, we use the homogeneous Poisson process (HPP) when the product is mature and the power process (PLP) when we observe growth (or decrease) in reliability (on the MTBF). Chapter 3 proposes to estimate the statistical properties (bias, coefficient of variation) of systems with maintenance on the basis of these processes.

Chapter 4 proposes to analyze the reliability of complex systems (series, parallel, k/n redundancy, series/parallel and parallel/series). Explicit solutions exist for maintenance-free applications, but few studies exist in the opposite case. The solutions then proposed systematically assume a constant failure rate, which is a realistic assumption for catastrophic failures and mature systems (Bayle 2021). To support our argument, we will propose two industrial applications for which Monte Carlo simulations will be used.

It is well known that temperature plays a major role in the reliability of components. Arrhenius proposed an empirical law to quantify its influence on the chemical reaction rate of sugarcane in the 19th century. In the 1940s, during the invention and development of the first transistors, Schockley was able to make the same observation and decided to use this same law to model the effect of temperature on the reliability of transistors. Currently, this law is widely used globally for the majority of components.

However, the Arrhenius law is valid only at a constant temperature, so for the majority of industrial applications, it is not directly applicable because the temperature experienced by operating systems is generally not constant. Sedyakin proposed a physical principle in reliability, allowing for the calculation of the survival function of a system undergoing, for example, temperature steps. The consequence of this principle is that the resulting probability law no longer follows a known law. Under certain industrial conditions, generally encountered for a majority of industrial applications, following Sedyakin's principle, it is possible to find an equivalent constant temperature in terms of reliability, which ultimately allows the use of the Arrhenius law.

Chapter 5 then proposes the use of Monte Carlo simulations to estimate the reliability of systems subjected to such conditions.

Chapter 6 offers potential assistance in sizing reliability tests. These may be necessary when a specific new technology component is introduced into a system without having relatively precise information on its reliability level. This may also be the case when a component used in an industrial application shows a reliability level well below that expected. In this case, on the basis of the failure analyses carried out to explain this reliability gap, a design or even manufacturing modification is proposed. The end user, or more often the system designer, requires proof that with this modification, the component will indeed have the expected reliability level. Only a reliability test can then demonstrate this.

There is nothing worse than carrying out this type of test and not being able to demonstrate it, generally due to failure. Thus, the size of a reliability test has a major influence. By means of a certain number of assumptions allowing the estimation of certain parameters of the chosen reliability model, by means of a certain number of industrial constraints (cost and duration of the test, number of components available, number of test means, etc.), it is possible by Monte Carlo simulations to carry out a "virtual" reliability test allowing for the optimization of certain input data of a reliability test, such as the levels of physical contributions that the components under test will undergo and the temperature levels to be applied, as much as possible.

Chapter 7 presents an industrial example where the inversion method for generating random data is not possible and for which the rejection method is used. We begin by detailing the reliability issue encountered, including the use of the decentered beta distribution as a reliability model, and present the principle of the rejection method. Next, we apply the maximum likelihood method to estimate the parameters of the decentered distribution, and then, we provide a justification for using this distribution rather than another.

Part 2 on *Maintainability* focuses on maintenance aspects and, more specifically, on estimating the number of failures required for the correct sizing of spare stocks. We develop the theoretical part and then return to the examples used for system reliability in Chapter 4 and present the results obtained by simulation in Chapter 8.

Part 3 focuses on *Availability*. Thus, Chapter 9 examines the influence of failure times and repair times on a system's availability. Repair implies maintenance actions; therefore, this chapter addresses only industrial applications with maintenance. Availability models are based on the theory of alternating renewal processes and give rise to complex equations based on convolution products. Of course, in certain special cases that we will examine, explicit solutions do exist.

For simplicity, the analytical solution uses the Laplace transform (the Laplace transform of a convolution product being equal to the product of the Laplace transforms), but the transition to the time domain requires the use of the inverse Laplace transform, which often does not yield explicit solutions. Here, again, Monte Carlo simulations allow us to overcome this impasse.

Part 4 focuses on *Safety*. Chapter 10 addresses the reliability of concurrent mechanisms at the component level and, more specifically, the distribution of their failure modes. This information is essential for analyzing failure modes and their effects on a system. For maintenance-free applications, explicit solutions exist, but for maintenance-based applications, which represent the majority of real-life cases, this is not possible, particularly for aging failure mechanisms, which are generally modeled by a Weibull distribution.

Chapter 11 proposes an analysis similar to the reliability and availability of systems but applied to the estimation of the probability of occurrence of a feared event (safety). In the same way, the assumption of an exponential law is systematically used so that, for failure mechanisms by aging systems with maintenance, explicit solutions of this probability are generally not available. We present an approximation based on Bernoulli's law with respect to certain conditions of testability of the systems, which allows us to find an explicit solution except for the aging mechanisms, where it uses the notion of Rocof, which does not have one. Here, again, Monte Carlo simulations potentially allow an estimation of this probability of occurrence. We will also see as a corollary that for certain industrial applications requiring very high levels of reliability, simulations do not always succeed in solving the problem. Very long simulation times and a very large number of simulations can be needed, which is incompatible with the computing power of the computers in the possession of most engineers.

PART 1

Reliability

1

Predictive Reliability

1.1. Concept of predictive reliability

Predictive reliability comes into play as soon as a product bill of material has been drawn up, the design files are sufficiently advanced and a certain number of tests have been carried out on prototypes. Its main objective is to check that the specifications in terms of reliability are being met and, of course, as we have seen in the foreword by Gilles Zwingelstein where he discusses operational safety (OS), to quantitatively provide input for maintenance (sizing of spare parts stocks), FMEA, testability and the various product fault trees.

To do this, a number of assumptions are made:

– Youth failures are not taken into account because they are filtered by the debugging process.

– Aging failures, except for certain subassemblies (fans, batteries, etc.), are not considered because they will not be visible during the operational life of the product. If this is the case, however, preventive maintenance actions or an estimate of the failure rate due to aging will then be taken into account for the final estimate.

– Premature aging of certain components due to a design weakness is not considered because it is determined by environmental robustness tests, derating analyses (compliance with a margin in relation to the design limit of a component) and worst case analyses, which allow the variability of the components to be taken into account, mechanical, thermal, electronic simulations, etc.

Thus, only the period of time when the product is mature (constant reliability) is considered in the predictive reliability analyses, as illustrated in the famous bathtub curve.

One possible method is the FIDES predictive reliability method. This is the method we use in this chapter.

Figure 1.1. *Illustration of the bathtub curve*

1.2. FIDES methodology

1.2.1. *Modeling reminder*

The objectives of the FIDES methodology (FIDES 2022) are, on the one hand, to enable a realistic assessment of the reliability of electronic products, including in systems that encounter harsh or very benign environments (storage), and, on the other hand, to provide a concrete tool for the construction and control of this reliability.

Its main characteristics are as follows:

– the existence of models for both electrical, electronic and electromechanical components as well as for electronic cards or certain subassemblies;

– the highlighting and consideration of all of the technological and physical factors that have an identified role in reliability;

– precise consideration of the life profile;

– taking into account accidental electrical, mechanical and thermal overloads (or overstress);

– taking into account failures linked to development, production, operation and maintenance processes;

– the ability to distinguish between multiple suppliers of the same component.

The FIDES reliability approach is based on the consideration of three components: technology, process and use. These components are considered for the entire lifecycle, from the product specification phase to the operation and maintenance phase.

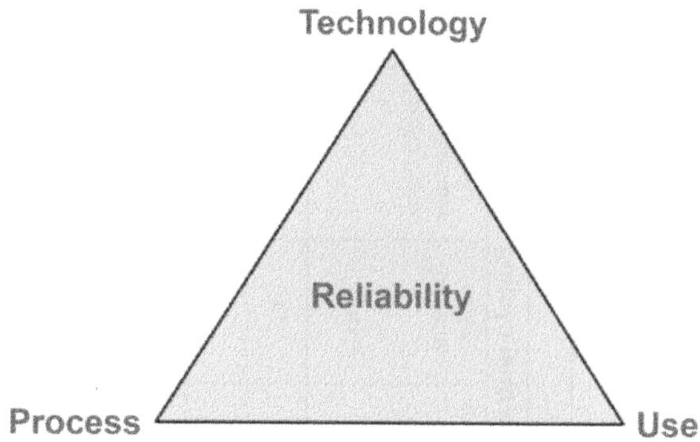

Figure 1.2. *Illustration of the pillars of the FIDES methodology. For a color version of this figure, see www.iste.co.uk/bayle/montecarlo.zip*

Technology covers both the article itself and its integration into the product. The *process* considers all practices and rules of the art from product specification to replacement. The use takes into account both the usage constraints defined by the product design and those in operation by the end user.

1.2.2. *Concept of life profile*

The creation of the life profile is a major step in estimating the product's predicted reliability and is often poorly specified and poorly analyzed. This is often because of the following factors:

– The proposed life profile is copied from one generation of equipment to another, which means that no one can justify the values of the parameters taken into account.

– The creation of the life profile is a joint effort between the client and the product designer. In fact, the client must specify the ambient external parameters that the product will be subjected to during its operational life, whereas the designer, on the basis of these data and the various characteristics of the product, will estimate its internal parameters.

Standard life profile									
		Temperature and humidity			Thermal cycling				Mechanics
Phase title	Calendar time (hours)	On/Off	Room temperatures (°C)	Humidity level (%)	ΔT (°C)	Number of cycles (/year)	Cycle duration (hours)	Maximal temperature during cycling (°C)	Random vibrations (Grms)
Off phase	5 110	Off	20	70	5	365	14	23	0.01
On phase	3 650	On	40	22	20	365	10	40	0.5

Table 1.1. *Example of a life profile*

– There is a great deal of variability in the levels of physical contribution. This is the case, for example, in the automotive sector, where the same type of car can have different life profiles depending on the type of driver, the type of road (motorway, city, countryside, mountain, etc.), the geographical location, the climatic conditions, etc.

Typically, predictive reliability estimates are made from constant values, leading to a single failure rate estimate, as illustrated in the example in Table 1.1 (FIDES 2022).

The values used are often average values, as in the example above, where the humidity level of 70% is a global average value. This approach has two drawbacks:

– the average value does not always represent an overall trend well, especially if the data are very dispersed;

– given that the laws of failure physics (Arrhenius, Coffin–Manson, etc.) are convex, an average value can then give imprecise results.

On the other hand, equating temperature variations (for example) can be an insurmountable task owing to different variabilities (day/night, seasonality, product use, etc.) that are difficult to model.

Therefore, Monte Carlo simulations can prove to be a more effective tool, particularly:

– a confidence interval is allowed for the estimated failure rate, which also makes it possible to respond, at least in principle, to certain comments made on this point (Pecht and Das 2020) in relation to the FIDES guide;

– by being able to estimate the relative importance of certain parameters of the life profile (temperature value or even of a particular phase of the profile considered), we can simplify it or pay particular attention to a given parameter.

1.3. Application example

Considering the following virtual example, the parameters Pi_Induit, Pi_Part and Pi_Process are normalized to 1.

Family	Technology family 1	Technology family 2	Technology family 3	Technology family 4	Pi Placement	Quantity
Capacitor	Aluminium capacitor	Liquid electrolyte			Analog function low level non interface	1
Capacitor	Ceramic capacitor	Type I	Category 2		Analog function low level non interface	1
Capacitor	Ceramic capacitor	Type II	non polymer terminaison		Analog function low level non interface	1
Capacitor	Ceramic capacitor	Type II	polymer terminaison	Category 2	Analog function low level non interface	1
Capacitor	Tantalum capacitors	Solid	CMS packaging		Analog function low level non interface	1
Discret semiconductor	Low power diode	Signal diode < 1A	SMD small signal L - lead plastic		Analog function low level non interface	1
Discret semiconductor	Low power diode	zener diode < 1.5W	SMD small signal L - lead plastic		Analog function low level interface	1
RF Discret semiconductor	Low power diode Si SiGe RF and HF		SMD small signal L - lead plastic		Analog function low level interface	1
Discret semiconductor	Power diode	zener diode > 1.5W	SMD small signal L - lead plastic		Analog power functions non interface	1
Discret semiconductor	Low power diode	Rectifying Diodes 1A to 3A	SMD small signal L - lead plastic		Analog function low level interface	1
Discret semiconductor	Power diode	Rectifying Diodes > 3A	SMD small signal L - lead plastic		Analog function low level non interface	1
Discret semiconductor	Low power transistor	Silicium bipolar < 5W	SMD small signal L - lead plastic		Analog function low level non interface	1
Discret semiconductor	Low power transistor	Silicium bipolar < 5W	SMD small signal L - lead plastic		Analog function low level non interface	1
Discret semiconductor	Low power transistor	Silicium MOS <5W	SMD small signal L-lead plastic		Analog function low level non interface	1
Fuse	CMS				Analog power function interface	1
Passive RF	Fixed				Analog function low level interface	1
Passive RF	Variables				Analog function low level interface	1
Piezoelectric component	Oscillator	SMD XO type			Analog function low level non interface	1
Connector	Coaxial	Soldered through hole			-	1
Connector	Circulars and rectangulars	Soldered through hole			-	1
Magnetic component	Multilayer inductor				Analog function low level non interface	1
Magnetic component	Transformator	Low power			Analog function low level non interface	1
Magnetic component	Transformator	Low power			Analog function low level non interface	1
Resistor	Resistive chip				Analog function low level non interface	1
Integrated circuit	Flash EEPROM EPROM memories	Numeric function none interface	TSOPI TSOPII		Numeric function non interface	1
Integrated circuit	Circuit Analog and Mixed circuits	Numeric function none interface	SSOP VSOP QSOP		Numeric function non interface	1
Integrated circuit	FGPGA CPLD FPGA Antifuse PAL	Numeric function none interface	PBGA BT		Numeric function non interface	1
Integrated circuit	Microprocessor Microcontrolor DSP	Numeric function none interface	PBGA CSP BT		Numeric function non interface	1
Optocouplor	Phototransistor	SO SOP SOL SOIC SOW			Analog function low level non interface	1
Printed circuit	210/210	Through holes			-	1

Table 1.2. *Example of a list of components based on FIDES*

1.3.1. *Classical estimation*

We obtain a failure rate of 8.3 Fits with the following distribution:

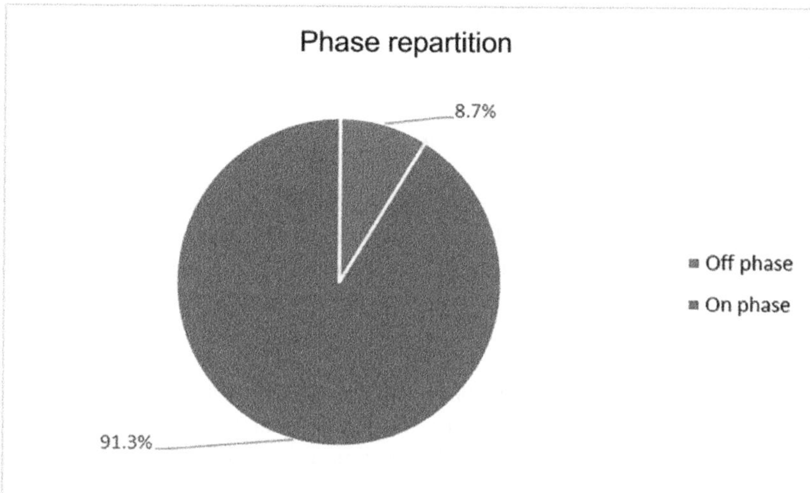

Figure 1.3. *Predicted failure rate repartition. For a color version of this figure, see www.iste.co.uk/bayle/montecarlo.zip*

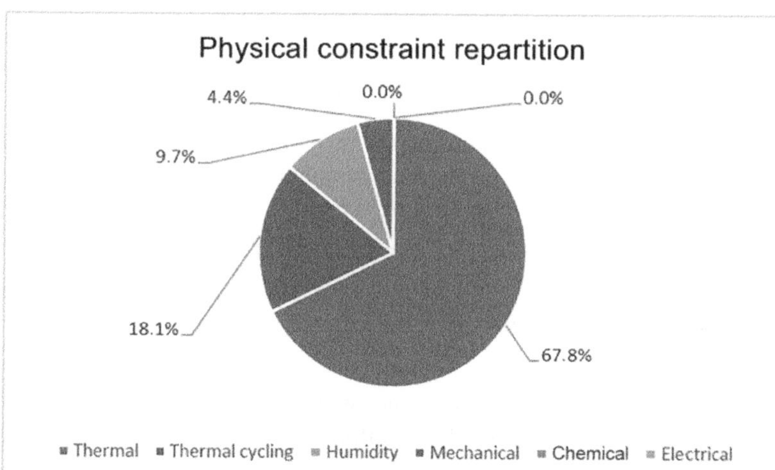

Figure 1.4. *Distribution of the predicted failure rate by physical contribution. For a color version of this figure, see www.iste.co.uk/bayle/montecarlo.zip*

The "Off" phase has a minor impact on the overall failure rate. It is therefore better to look at the sensitivity of the "On" phase parameters.

The physical, chemical and electrical contributions have no effect, whereas the thermal contribution represents more than two-thirds of the overall failure rate. A sensitivity analysis is then necessary for the temperature contribution.

1.3.2. *Estimation with a uniform law*

Instead of considering a single value for the levels of physical contributions, we propose here to have an interval of values following a uniform law and, in particular, to apply it to the thermal contribution. Thus, instead of considering a temperature of 40°C for the "On" phase, we propose taking into account a uniform law of parameters (35°C; 45°C).

From a confidence level of 90%, we obtain a failure rate of 8.4 Fits with a confidence interval of (6.96; 10.12) Fits. We carried out 1,000 simulations, and the failure rate was obtained by taking the arithmetic mean of the 1,000 simulated failure rates. From these results and sorting them in ascending order, we obtain, for a confidence level of 90%, the lower bound at the 50th rank and the upper bound at the 950th rank. If we had chosen a uniform law with the parameter (20°C; 60°C), we would have obtained an average failure rate of 9.98 Fits with a confidence interval of (4.67; 20.19) Fits.

REMARK.–

These simulation results indicate the following:

– The average value of the failure rate is higher than the "classical" value, especially since the extent of the normal distribution is large. Importantly, the uniform distribution $U[\theta a, \theta b]$ with $\theta b > \theta a$ is involved in the Arrhenius distribution in the following form: $\lambda(\theta) = C.\exp\left(\frac{-\text{Ea}}{\text{Kb}.U[\theta a,\theta b]}\right)$. The activation energy represents the sensitivity of the failure rate to temperature θ, and the failure rate will be much more strongly impacted for temperature θb than for θa because of the convexity of the exponential function.

– We can also notice the same trend with the confidence interval, which "widens" all the more as the extent of the uniform law is important.

These two phenomena are more important as the activation energy is high.

1.3.3. *Estimation with a normal distribution*

However, since the simulated temperature values are all equiprobable, this is not very realistic because the probability of having extreme values in the considered interval is rather low. Additionally, we can replace the uniform law with a normal law. However, this approach requires more information on the temperature values than the previous approach does.

Let us therefore consider a normal distribution with parameters N (40°C; 5°C). From a confidence level of 90%, we obtain an average failure rate of 8.6 Fits with a confidence interval of (6.15; 12.3) Fits.

However, the normal law has two drawbacks in the area of reliability:

– Its symmetry about the mean is because there is no physical reason for this to be the case.

– Its definition ranges from $-\infty$ to $+\infty$, which generally poses a problem, particularly for humidity levels.

To avoid the first drawback, we can use the Weibull law, which can be very asymmetrical depending on the value of the shape parameter. For the second drawback, for example, for the humidity rate, we can use a beta law whose domain of definition is [0; 1], which is therefore in line with the humidity rate, which can vary from 0 to 100%.

1.4. Maintaining a reliability specification

One of the goals of predictive reliability is to verify compliance with a specification. Typically, this is achieved by making a "brutal" comparison between the reliability estimate and the allocated target.

For example, taking the previous example where we had an objective failure rate of 10 Fits, the classic analysis leads to this objective being met since the estimated failure rate is lower (8.4 Fits). However, is this really the case? In other words, is the fact that it is lower than the objective sufficient to conclude?

The advantage of taking into account uncertainties in the life profile allows us to obtain an estimate but also a confidence interval. Thus, to have good confidence in the holding of the specification, we will rather compare the objective to the upper bound of the estimated failure rate. In the previous section, we had a bilateral confidence interval (lower and upper bounds). Here, we will choose a unilateral upper bound, that is, the 900th rank (90% confidence level for 1,000 simulations).

If we take the example with a uniform law on the temperature of the "On" phase, we obtain a unilateral upper bound of 9.9 Fits, which is very close to the objective.

NOTE.–

Generally, the reliability objective is expressed in terms of the following:

– Probability of failure for a given time Tobj and confidence level for maintenance-free applications such as space, nuclear or military. The probability of failure and the failure rate are, for an exponential distribution in accordance with the assumption of section 1.1, linked by the following relationship:

$$F(Tobj) = 1 - \exp(-\lambda.Tobj)$$

– MTBF for most applications that have maintenance actions. The failure rate and MTBF are related by the following relationship:

$$MTBF = \frac{1}{\lambda}$$

1.5. Summary

– Estimating the predicted reliability of a system is usually essential during its design phase.

– The life profile that the system will undergo during its operation has a major influence on the estimation of the predicted reliability.

– The levels of physical contribution of the life profile considered are often very dispersed over time, whereas generally, the predictive reliability is estimated from average levels.

– For greater realism, we propose replacing, when necessary, the average levels with probability laws, based on the information collected, in the form of Monte Carlo simulations.

– The advantage of these simulations is that they can estimate the importance of certain phases of the life profile and therefore the relevance of the assumptions made on the levels of corresponding physical contributions (temperature, vibration, humidity).

2

Statistical Characteristics of Exponential and Weibull Distributions

2.1. Refresher about exponential and Weibull distributions

2.1.1. *Exponential distribution*

For an exponential distribution with parameter λ, the probability density is given by: $f_e(t) = \lambda.\exp(-\lambda.t)$.

Its reliability function is therefore equal to:

$$R_e(t) = \int_0^t \lambda.\exp(-\lambda.u).\mathrm{du} = \exp(-\lambda.t)$$

The MTTF is given by:

$$MTTF_e = \int_0^{+\infty} R_e(t).dt = \int_0^{+\infty} \exp(-\lambda.t).dt = \frac{1}{\lambda}$$

2.1.2. *Weibull distribution*

For a Weibull distribution of parameters η and β, the probability density is given by:

$$f_w(t) = \left(\frac{\beta}{\eta}\right).\left(\frac{t}{\eta}\right)^{\beta-1}.exp\left(-\left(\frac{t}{\eta}\right)^{\beta}\right)$$

Its reliability function is therefore equal to $R_w(t) = exp\left(-\left(\frac{t}{\eta}\right)^{\beta}\right)$.

The MTTF is given by:

$$MTTF_w = \eta.\Gamma\left(1+\frac{1}{\beta}\right) \text{ where } \Gamma \text{ represents the gamma function}$$

The Weibull distribution MTTF is proportional to the scaling parameter η but does not depend much on the shape parameter β when the latter is greater than or equal to 1, which translates to aging mechanisms because the function $\Gamma\left(1+\frac{1}{\beta}\right)$, for $\beta > 1$, can vary between only ~ 0.89 and 1. Therefore, the statistical properties of the parameter β barely affect the statistical properties of the MTTF. On the other hand, those of the parameter η will have a direct impact.

2.2. Parameter estimation for a reliability model using the maximum likelihood method

Let us assume that we have observed "n" failures at times t_1, t_2,..., T_n. Let T_1, T_2,..., T_n denote the failure times corresponding to the model adopted. The likelihood consists of estimating the probability P that the theoretical times T_i correspond to the observed times t_i. The likelihood L is then written as:

$$\mathrm{L} = P[t_1 = T_1] \cap P[t_2 = T_2] \cap \ldots \cap P[t_n = T_n]$$

Since these events are independent, we obtain:

$$\mathrm{L} = \prod_{i=1}^{n} P[t_i = T_i]$$

To simplify the calculations, it is preferable to use the loglikelihood Λ, that is:

$$\Lambda = \ln(L) = \sum_{i=1}^{n} ln(P[t_i = T_i]) \qquad [2.1]$$

2.2.1. Full data

By full data, we mean that all of the components have failed testing. On the other hand, if we assume that the failure is instantaneously detected (at least in a negligible time compared with the failure instants), we can write that:

$$\Lambda = ln(L) = \sum_{i=1}^{n} ln(f(t_i))$$

2.2.2. *Type II right-censored data*

After completing the reliability test or an operational observation, not all of the components may fail. Let us assume that in the "n" components being tested, only "r" failed.

The remaining "n-r" components are censored. In this case, we obtain, if we denote by T_r, the time of the last failure:

$$L = \sum_{i=1}^{r} f(t_i). \prod_{j=1}^{n-r} R(T_r)$$

Therefore, the log-likelihood is written as:

$$\Lambda = \ln(L) = \sum_{i=1}^{r} ln\big(f(t_i)\big) + (n-r). ln\big(R(T_r)\big) \qquad [2.2]$$

2.2.2.1. *Exponential distribution log-likelihood for type II right-censored data*

In this case, we have $f(t) = \lambda. \exp(-\lambda. t)$ and $R(t) = \exp(-\lambda. t)$, from which

$$\Lambda = \sum_{i=1}^{r} [ln(\lambda) - \lambda. t_i] + (n-r). ln\big(\exp(-\lambda. T_r)\big)$$

That is,

$$\Lambda = \text{r}. \ln(\lambda) - \lambda. \sum_{i=1}^{r} t_i - \lambda. (n-r). T_r$$

The derivative of the log-likelihood with respect to λ is then equal to:

$$\frac{\delta\Lambda}{\delta\lambda} = \frac{r}{\lambda} - \sum_{i=1}^{r} t_i - (n-r). T_r$$

If we look for the value of λ for which this derivative cancels out, we obtain:

$$\hat{\lambda} = \frac{\text{r}}{\sum_{i=1}^{r} t_i + (n-r). T_r} \qquad [2.3]$$

2.2.2.2. *Weibull distribution log-likelihood for type II right-censored data*

The log-likelihood is written as:

$$\Lambda = \ln(L) = \sum_{i=1}^{r} ln\left(\left(\frac{\beta}{\eta}\right).\left(\frac{t_i}{\eta}\right)^{\beta-1}.exp\left(-\left(\frac{t_i}{\eta}\right)^{\beta}\right)\right) + (n-r).ln\left(exp\left(-\left(\frac{T_r}{\eta}\right)^{\beta}\right)\right)$$

that is:

$$\Lambda = \mathrm{r}.\ln(\beta) - \mathrm{r}.\ln(\eta) + (\beta-1).\sum_{i=1}^{r} ln\left(\frac{t_i}{\eta}\right) - \sum_{i=1}^{r}\left(\frac{t_i}{\eta}\right)^{\beta} - (n-r).\left(\frac{T_r}{\eta}\right)^{\beta}$$

that is:

$$\Lambda = \mathrm{r}.\ln(\beta) - \mathrm{r}.\upbeta.\ln(\eta) + (\beta-1).\sum_{i=1}^{r} ln(t_i) - \frac{1}{\eta^{\beta}}.\left(\sum_{i=1}^{r} t_i{}^{\beta} + (n-r).T_r{}^{\beta}\right)$$

To maximize the log-likelihood, this equation must be differentiated with respect to η and β, that is:

$$\frac{\partial\Lambda}{\partial\upeta} = \frac{-\beta.r}{\eta} + \frac{\beta}{\eta^{\beta+1}}.\left(\sum_{i=1}^{r} t_i{}^{\beta} + (n-r).T_r{}^{\beta}\right)$$

This derivative is then set to 0, from which:

$$\hat{\eta} = \left(\frac{\sum_{i=1}^{r} t_i{}^{\hat{\beta}} + (n-r).T_r{}^{\hat{\beta}}}{r}\right)^{1/\hat{\beta}}$$

On the other hand, we have:

$$\frac{\partial\Lambda}{\partial\upbeta} = \frac{r}{\beta} - r.ln(\eta) + \sum_{i=1}^{r} ln(t_i) - \frac{1}{\eta^{\beta}}.\left(\sum_{i=1}^{r} t_i{}^{\beta}.ln(t_i) + (n-r).T_r{}^{\beta}.ln(T_r) - ln(\eta).\left(\sum_{i=1}^{r} t_i{}^{\beta} + (n-r).T_r{}^{\beta}\right)\right)$$

that is:

$$\frac{\partial\Lambda}{\partial\beta}=\frac{r}{\beta}+\sum_{i=1}^{r}ln(t_i)-\frac{1}{\eta^{\beta}}.\left(\sum_{i=1}^{r}TTF_i^{\ \beta}.ln(t_i)+(n-r).T_r^{\ \beta}.ln(T_r)\right)$$

Because the term $-r.ln(\eta)+\frac{1}{\eta^{\beta}}.\left(ln(\eta).\left(\sum_{i=1}^{r}t_i^{\ \beta}+(n-r).T_r^{\ \beta}\right)\right)$ is zero for $\eta=\hat{\eta}$, finally, by replacing η with its estimator:

$$\frac{r}{\hat{\beta}}+\sum_{i=1}^{r}ln(t_i)-r.\left(\frac{\sum_{i=1}^{r}t_i^{\ \hat{\beta}}.ln(t_i)+(n-r).T_r^{\ \hat{\beta}}.ln(T_r)}{\sum_{i=1}^{r}t_i^{\ \hat{\beta}}+(n-r).T_r^{\ \hat{\beta}}}\right)=0$$

This equation can be solved via numerical methods only because it is transcendental (it involves both x and ln(x)). As such, the fact that the estimator of the parameter β does not have explicit expressions makes it possible to determine its properties (bias and coefficient of variation)

2.3. Estimator properties

2.3.1. *Notion of bias*

We would like the average of the realizations of an estimator to be equal to the true value θ of the parameter.

If so, the estimator is said to be unbiased. It is defined from a theoretical point of view as follows:

$$Bias(\hat{\theta})=\mathrm{E}(\hat{\theta})-\theta$$

where:

– θ represents a parameter of the model under consideration;

– $\hat{\theta}$ represents its estimate;

– E represents the mathematical expectation (theoretical average).

From a simulation perspective, it is defined as follows:

$$Bias(\hat{\theta})=\mathrm{Mean}(\hat{\theta})-\theta \qquad [2.4]$$

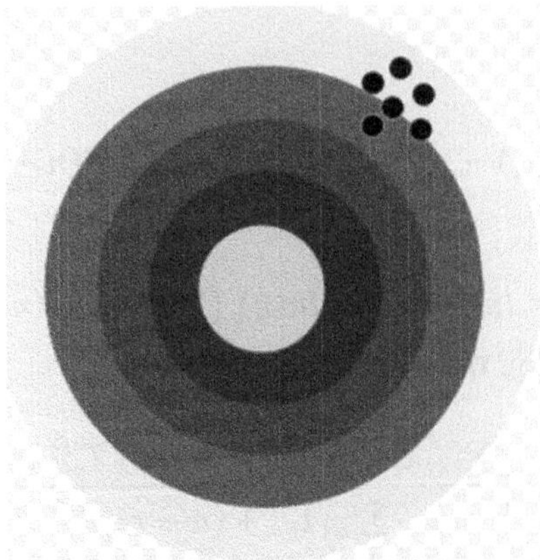

Figure 2.1. *Illustration of a biased element. For a color version of this figure, see www.iste.co.uk/bayle/montecarlo.zip*

Figure 2.2. *Illustration of an unbiased element. For a color version of this figure, see www.iste.co.uk/bayle/montecarlo.zip*

Generally, the expression of the bias of a parameter estimator does not have analytical solutions. This bias can be expressed only in a few situations, such as with the exponential distribution below. To better compare the simulation results, it is preferable to use the notion of relative bias. This is defined by:

$$Bias(\hat{\theta}) = \frac{\theta - \mathrm{Mean}(\hat{\theta})}{\theta} \qquad [2.5]$$

2.3.2. *Notion of coefficient of variation CV*

Another statistical property of estimators is variance, which represents the dispersion of the estimates of a parameter in a reliability model. This is defined by:

$$Var(X) = \mathrm{E}[(X - E[X])^2]$$

which can be found in the following much easier-to-use form:

$$Var(X) = \mathrm{E}[X^2] - (E[X])^2$$

To obtain the same unit as the bias, one uses the standard deviation defined by:

$$\sigma(X) = \sqrt{Var(X)}$$

or still:

$$\sigma(X) = \sqrt{\mathrm{E}[X^2] - (E[X])^2}$$

The disadvantage of this definition is that it is an absolute quantity, and to be homogeneous with the use of the relative bias previously defined, we prefer to use the notion of the coefficient of variation (CV), which is defined as follows:

$$CV(X) = \frac{\sigma(X)}{E[X]}$$

2.4. Simulation of failure times by inverting the probability of failure

The inversion method consists of using the reciprocal function (quantile) of the probability of failure F(t), which, by definition, lies between 0 and 1.

2.4.1. *Generating failure times for the exponential distribution*

The probability of failure is written as:

$$F(t) = 1 - \exp(-\lambda . t)$$

Consequently, its reciprocal function F^{-1} is written as:

$$F^{-1}(p) = \frac{-ln(1-p)}{\lambda} \qquad [2.6]$$

If $F(t)$ is a probability as a function of time, $F^{-1}(p)$ is the time after which "p%" failures have been observed. It can be shown (Nikulin et al. 2007) that the statistic $F(T)$ follows a uniform distribution over the interval [0,1]. Consequently, to generate TTF failure times randomly, one should use the following equation:

$$TTF = \frac{-ln(1-U[0,1])}{\lambda} \quad [2.7]$$

2.4.2. *Generating failure times for the Weibull distribution*

Here, the probability of failure is written as:

$$F(t) = 1 - \exp\left(-\left(\frac{t}{\eta}\right)^{\beta}\right)$$

from which:

$$TTF = \eta.\left(ln\left(\frac{1}{1-U[0,1]}\right)\right)^{1/\beta} \quad [2.8]$$

We use this method only for the Weibull distribution since for the exponential distribution, we have an analytic expression of the bias and variance on the parameter λ. To generate failure times according to a Weibull distribution, we use the inversion method with the probability of failure defined by equation [2.8]. Firstly, we consider a Weibull distribution of parameters η = 1,000 and β = 3 and generate 10 failure times via simulation. From the maximum likelihood method, we obtain the following Weibull plot shown in Figure 2.3.

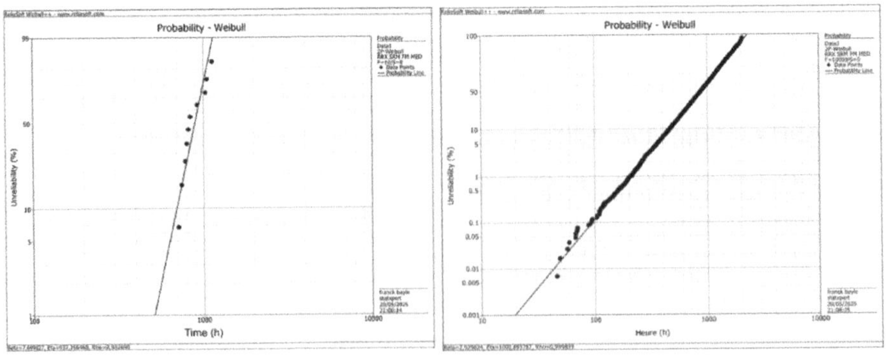

Figure 2.3. *Weibull unreliability simulations for 10 and 1,000 failure times*

The blue dots represent the failure times generated by the simulation, whereas the line represents the probability of failure obtained from the estimation of the parameters of the Weibull distribution. Although we started from the theoretical values of these parameters, they do not all belong to the model considered for 10 failures, which at first sight may be surprising. This is because of the random aspect of reliability, especially for small numbers of failures. If we compare the theoretical probability density with the histogram of the failure times and compare the respective figures for 10 (left) and 10,000 failures (right), we obtain the result shown in Figure 2.4.

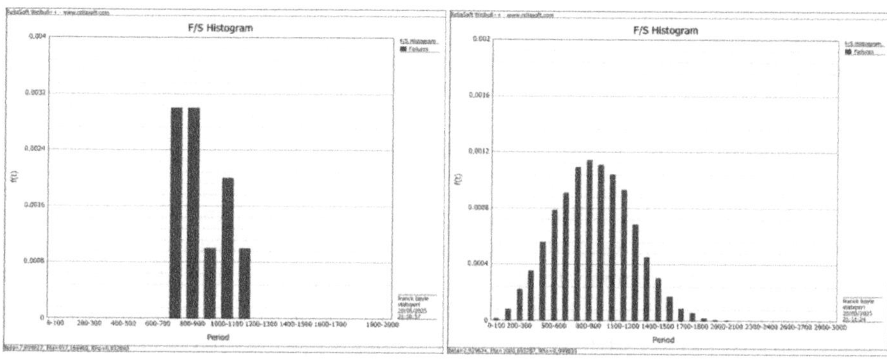

Figure 2.4. *Comparison between 10 and 10,000 failures simulation. For a color version of this figure, see www.iste.co.uk/bayle/montecarlo.zip*

Although the histogram and probability density are similar for 10,000 failures, they are not comparable for 10 failures. We can deduce that, for small numbers of failures, bias and the coefficient of variation may not be very good.

2.5. Impact of temperature

Most components exhibit failure mechanisms that are due to temperature. As such, the Weibull model is no longer sufficient. A family of models widely employed for reliability are AFT models. They assume that the physical contribution (temperature in this case) does not modify the shape of the distribution (the shape parameter β is independent of temperature) but only its scale parameter (the scale parameter η depends on temperature θ). Therefore, the probability of failure can be written as:

$$F(t,\theta) = 1 - \exp\left(-\left(\frac{t}{\eta(\theta)}\right)^{\beta}\right)$$

For temperature, the Arrhenius law is generally utilized such that:

$$F(t,\theta) = 1 - \exp\left(-\left(\frac{t}{C.exp\left(\frac{Ea}{Kb.\theta}\right)}\right)^{\beta}\right) \quad [2.9]$$

where:

– C is a pre-exponential constant, which represents the rate of the fastest reaction that can take place when the temperature is very high;

– Ea is the activation energy;

– Kb is the Boltzmann constant.

2.6. Relative bias and coefficient of variation of the Weibull parameters

We observed that the estimator of a parameter is a random value. As such, a single simulation cannot be used to estimate a parameter. Furthermore, to minimize the inaccuracy of the computed values and obtain simulation times that are not too long, we choose Nb_Simu = 1,000 simulations. We start with a scale parameter of the Weibull distribution such as η = 1,000, as this value will have no influence on the results that will be obtained.

We then vary the number of failures "n" (7, 10, 20, 40, 70 and 100) as well as the value of the shape parameter β (1.5, 2, 4, 7 and 10). We then obtain the following results.

2.6.1. Relative bias of β

The bias of the shape parameter β of a Weibull distribution depends on the number of failures but not on its value, explaining why the different curves are superimposed. As we previously observed, if we look into the MTTF, this will not have much influence on its estimate. To this end, we define the relative error of the MTTF according to a tolerance of β expressed in percentage, that is:

$$Err(\beta; Bias) = \frac{\Gamma\left(1+\frac{1}{\beta}\right) - \Gamma\left(1+\frac{1}{\beta.(1+bias)}\right)}{\Gamma\left(1+\frac{1}{\beta}\right)}$$

We then obtain Figure 2.6.

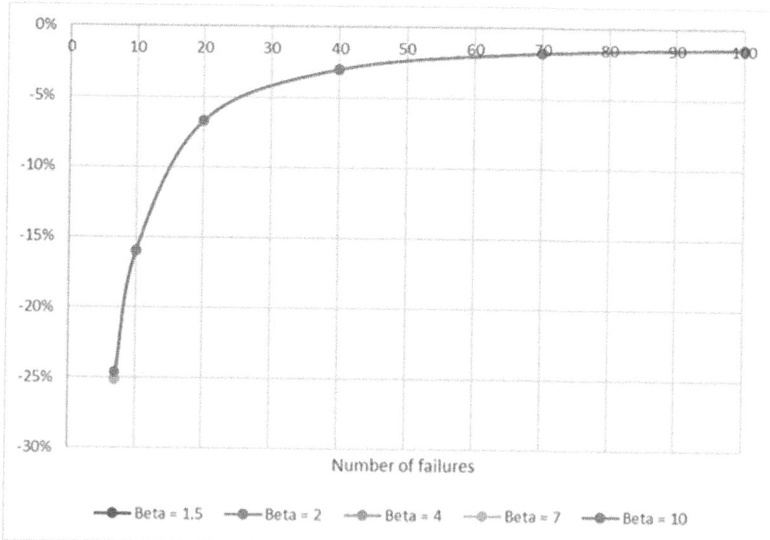

Figure 2.5. *Relative bias of the shape parameter β. For a color version of this figure, see www.iste.co.uk/bayle/montecarlo.zip*

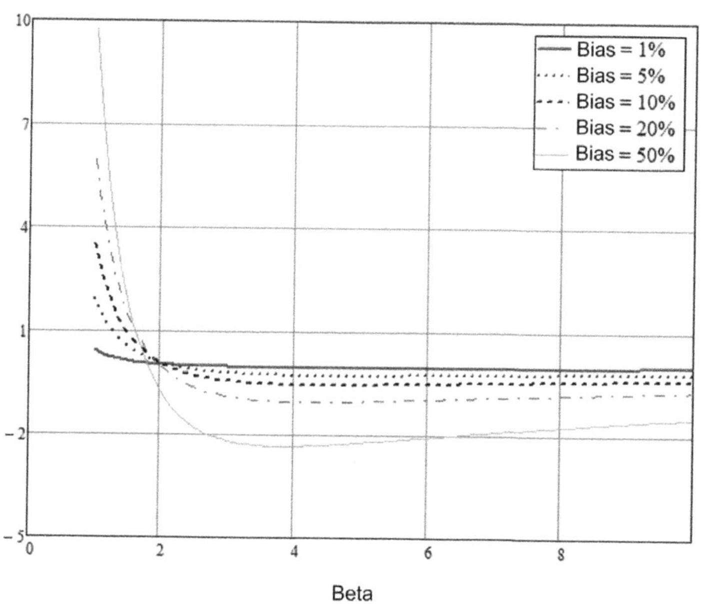

Figure 2.6. *Influence of the variability of β on the Weibull distribution MTTF. For a color version of this figure, see www.iste.co.uk/bayle/montecarlo.zip*

On the other hand, if we consider the probability of failure after a given mission time, it can have a significant impact.

2.6.2. *Relative bias of the parameter η*

The bias of the parameter η is small and virtually independent of the number of failures and the value of the parameter β. This is therefore also true for the MTTF.

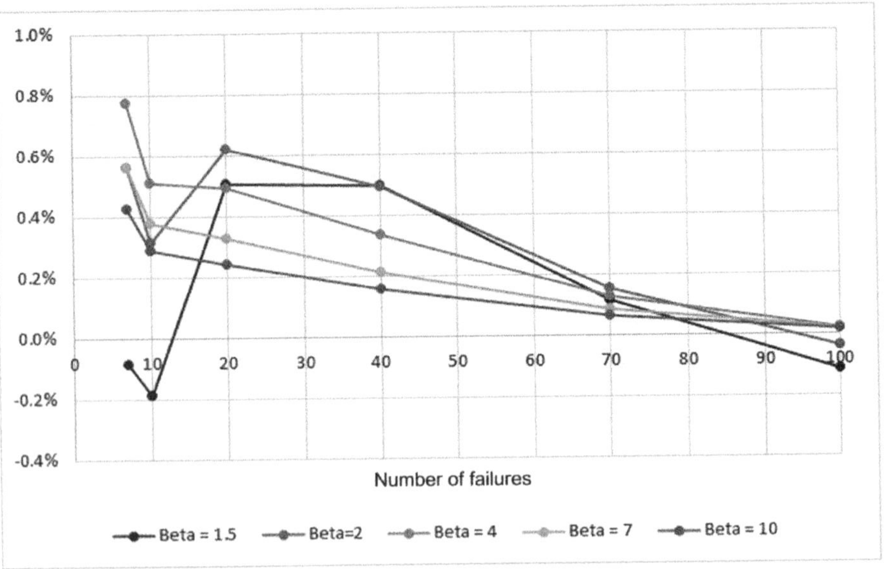

Figure 2.7. *Relative bias of the parameter η. For a color version of this figure, see www.iste.co.uk/bayle/montecarlo.zip*

The parameter η is barely biased, whether by the number of failures or the parameter β.

2.6.3. *Coefficient of variation of the parameter β*

The coefficient of variation of the parameter β depends on the number of failures but not much on its value.

It remains significant for a small number of components.

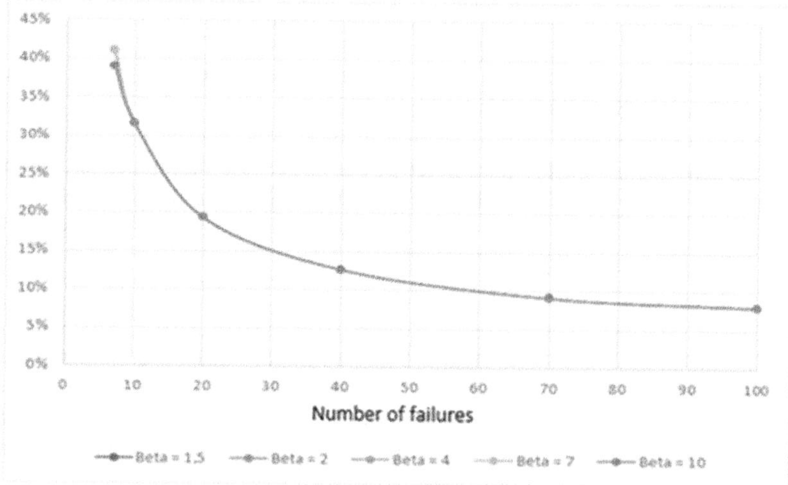

Figure 2.8. *Coefficient of variation of the parameter β. For a color version of this figure, see www.iste.co.uk/bayle/montecarlo.zip*

2.6.4. *Coefficient of variation of the parameter η*

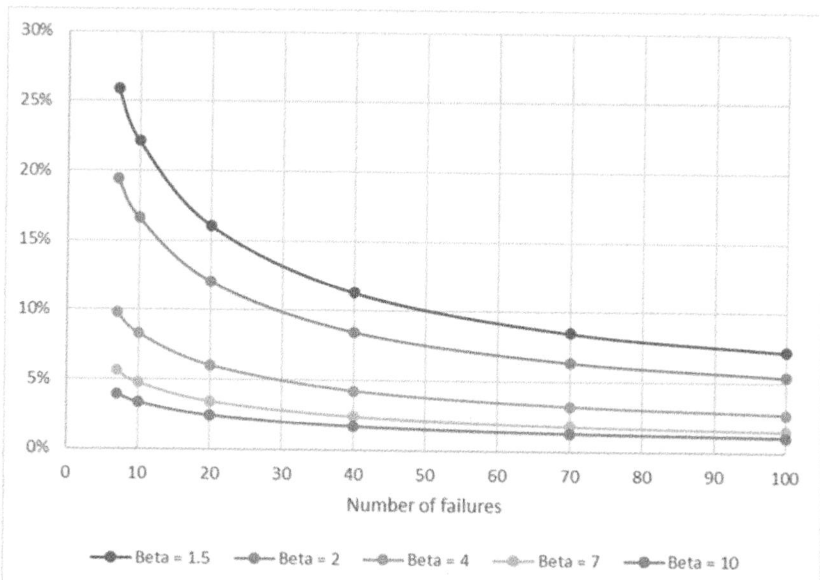

Figure 2.9. *Coefficient of variation of the parameter η. For a color version of this figure, see www.iste.co.uk/bayle/montecarlo.zip*

The coefficient of variation of the parameter η depends on the number of failures but also on the parameter β. For the dependency on the number of failures, the result makes sense since the more information we have (the number of failures), the less uncertainty there is about what it is being said.

For the dependency on the parameter β, the larger this parameter is, the greater the decrease in dispersion. This is because, for large values of the parameter β, all of the components tend to behave in the same way, which decreases the dispersion of the estimator of the parameter β.

2.7. Simulation scenarios considered from the different parameters

In addition to analyzing the influence of the parameters β, Ea and C, we also wanted to determine the influence of the number of temperature points as well as the spacing between the temperature values.

For this purpose, we consider Monte Carlo simulations with two different temperatures and spacings with two batches of components. All of the components are assumed to be independent and identically distributed according to an Arrhenius–Weibull type of AFT model.

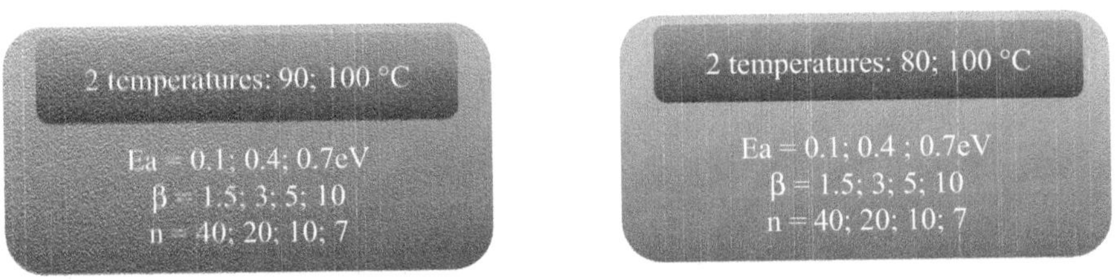

Figure 2.10. *Different simulation scenarios considered. For a color version of this figure, see www.iste.co.uk/bayle/montecarlo.zip*

2.7.1. *Relative bias of β*

The relative bias on β depends only on the number of "n" components. Accordingly, the larger n is almost independent of Ea and β, as illustrated in Figure 2.11.

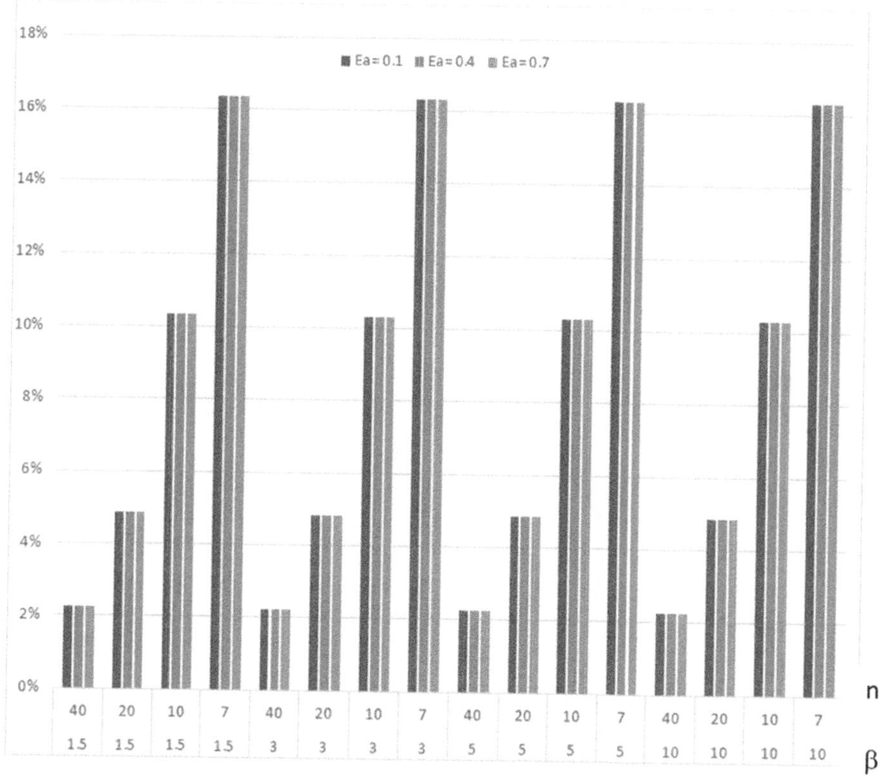

Figure 2.11. *Relative bias on the parameter β. For a color version of this figure, see www.iste.co.uk/bayle/montecarlo.zip*

2.7.2. *Bias of Ea*

The relative bias of Ea is small, especially for larger values of Ea, as shown in Figure 2.12.

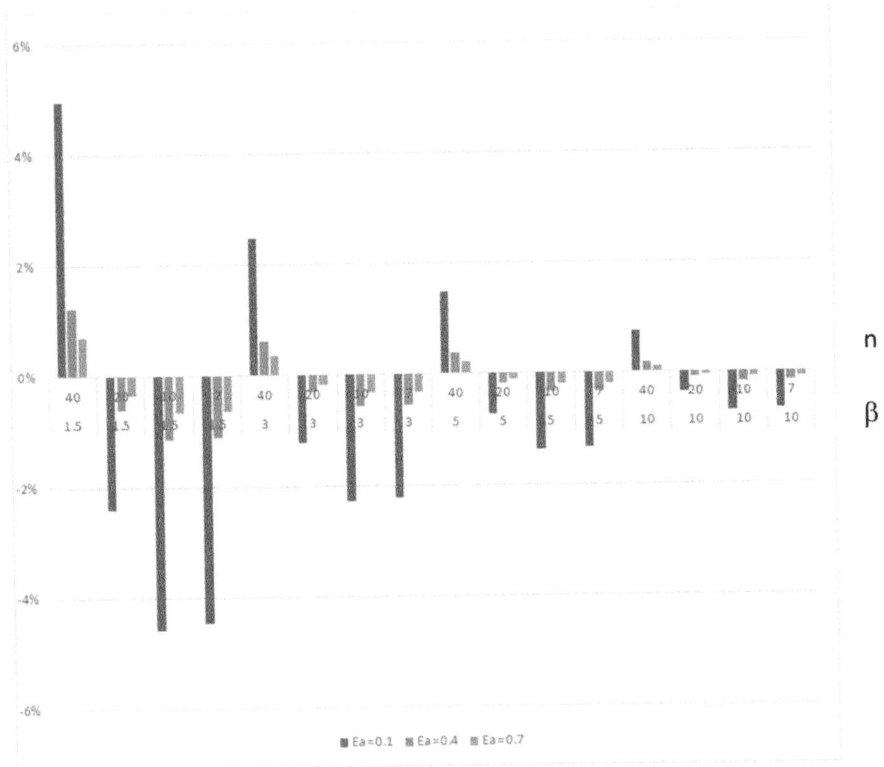

Figure 2.12. *Relative bias for parameter Ea. For a color version of this figure, see www.iste.co.uk/bayle/montecarlo.zip*

2.7.3. *Bias of C*

The relative bias of C is very significant, especially since the value of β is small and Ea is large.

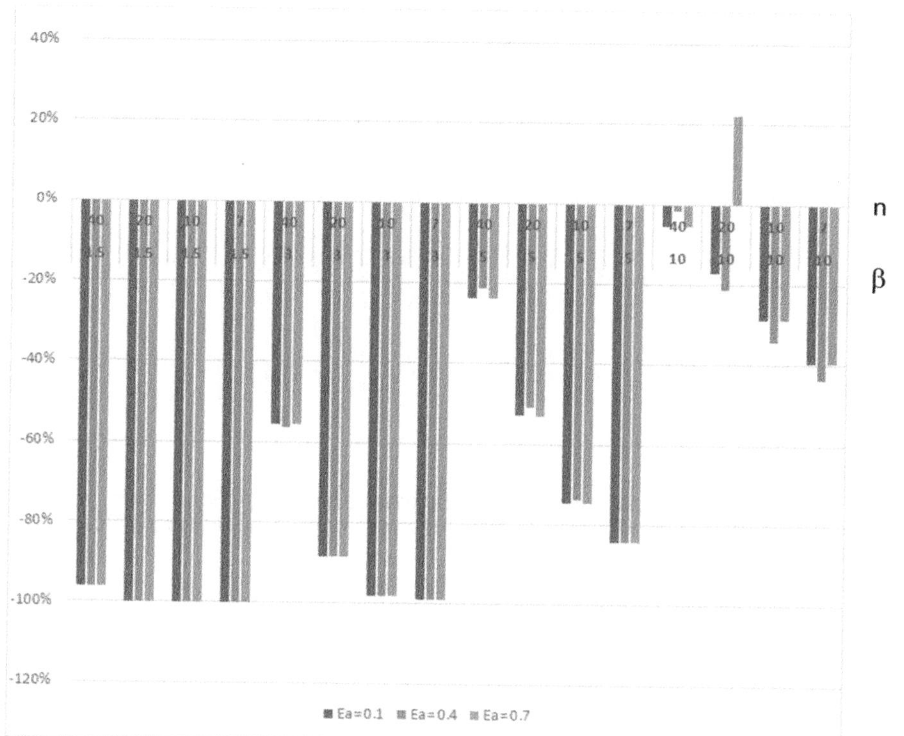

Figure 2.13. *Relative bias for parameter C. For a color version of this figure, see www.iste.co.uk/bayle/montecarlo.zip*

2.7.4. Coefficient of variation of β

The coefficient of variation of β depends mainly on the number of components. However, it remains quite small.

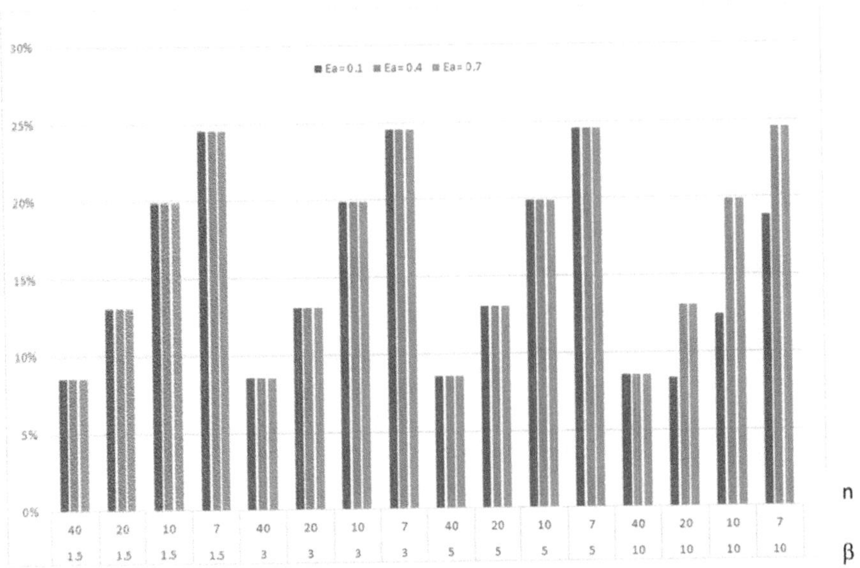

Figure 2.14. *Coefficient of variation with parameter β. For a color version of this figure, see www.iste.co.uk/bayle/montecarlo.zip*

2.7.5. *Coefficient of variation with Ea*

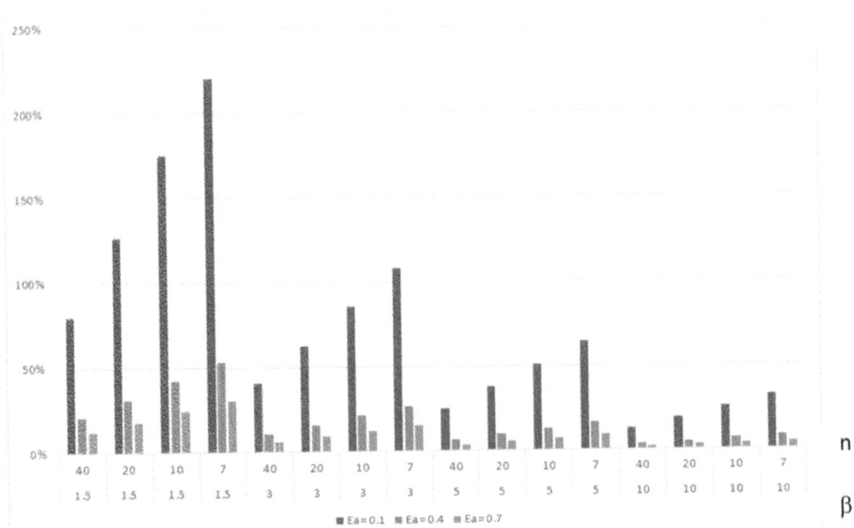

Figure 2.15. *Coefficient of variation with parameter Ea. For a color version of this figure, see www.iste.co.uk/bayle/montecarlo.zip*

The coefficient of variation with Ea depends not only on the number of components but also on the activation energy Ea itself.

2.7.6. *Coefficient of variation with C*

The coefficient of variation with C depends on the number of components but mainly on the activation energy Ea itself.

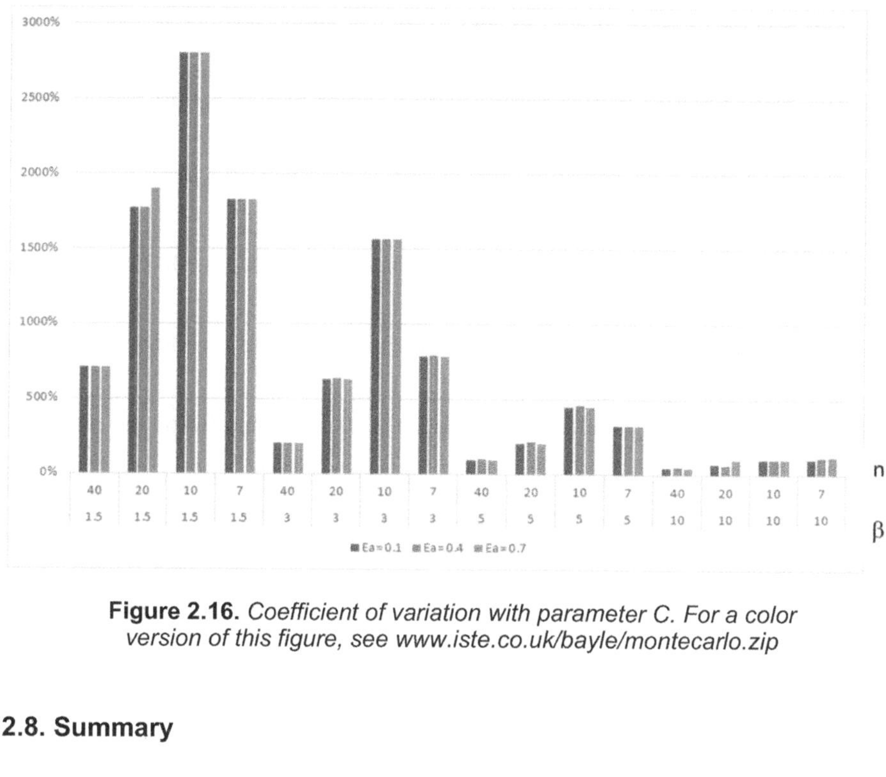

Figure 2.16. *Coefficient of variation with parameter C. For a color version of this figure, see www.iste.co.uk/bayle/montecarlo.zip*

2.8. Summary

– Bias and variance are the two statistical properties of parameter estimators in a reliability model.

– To better compare the results obtained, we used the relative bias and the coefficient of variation.

– The relative bias of the shape parameter β can be significant if the number of failures is small. It does not depend on the values of parameters η and β. The maximum likelihood method thus tends to overestimate the value of β.

– The relative bias of the scale parameter η of a Weibull distribution is negligible, regardless of the number of failures and the value of the parameter β.

– The coefficient of variation of parameter β can be significant if the number of failures is small. It does not depend on the values of parameters η and β.

– The coefficient of variation of parameter η can be significant because it depends not only on the number of failures but also on the value of the parameter β. This value is greater for larger β.

– The relative bias of the parameter β is not very sensitive to temperature (in fact, to the activation energy).

– The relative bias of the activation energy is small, regardless of the number of failures and the values of Ea and β.

– The relative bias of parameter C is significant in all the cases. It is relatively reliable for many failures (> 40) and a high value of β (> 5).

– The coefficient of variation of the parameter β is quite small and does not depend on temperature.

– The coefficient of variation of the parameter Ea can be very high, especially for small values of Ea.

– The coefficient of variation of parameter C is very significant but does not depend on Ea.

3

System Reliability

The principles of system reliability are well known, whether for serial systems, parallel systems, "k/n" active redundancy, reliability diagrams, etc. However, it should be observed that the theoretical elements that describe them are generally developed for applications in which there are no maintenance mechanisms. In other words, if a system element fails, it remains in that state until the system is fully operational. It is clear that this is not usually the case during the operational life of a system. We can therefore count two main categories of systems:

– Those *that do not have maintenance*, such as those embedded in a rocket and a military weapon. Here, it is the probability of succeeding in the mission that matters, and since it must be high, virtually no failures can be admitted.

– Those that *have maintenance* and represent most industrial applications (aeronautics, automobiles, railways, household appliances, hi-fi, computers, vacuum cleaners, etc.). The problem is that the theory found in the literature, which concerns maintenance-free systems, cannot be used in this context. Therefore, the term failure rate should not be employed but rather the failure intensity or failure occurrence rate (Rocof), as defined in section 3.3.

Given that technology and customs evolve, dealing with system renewal increasingly becomes the main concern. A telephone, a television, a vacuum cleaner, etc., will no longer be repaired (which is a maintenance action) but will be replaced with a new one for two essential reasons:

– repairing can cost as much or even more than a new system;

– as technology evolves, the new generation of systems is more attractive to the user than the previous generation.

For a color version of all of the figures in this chapter, see www.iste.co.uk/bayle/montecarlo.zip.

This chapter therefore aims to cover this theoretical lack as well as possible. To this end, we propose two maintenance strategies respectively at element and system level. In any case, the maintenance will be considered perfect (with a negligible repair time).

3.1. Assumptions

We look into a system mounted on a carrier with the following characteristics:

– the carrier has an operational activity that is the same every day (On/Off setting, duration of operational phases, levels of physical contributions, etc.);

– it is assumed that temperature is the most predominant physical contribution. The reliability level of the system remains unchanged when the equipment is not powered (Off);

– before each operational phase, when the system is powered up, its state (functional or faulty) is known and it is assumed that the coverage of the tests is 100%, although in practice, this rate is lower (such as ~ 90% in aeronautics);

– it is assumed that the system comprises "n" elements and that these are independent of each other;

– we do not consider multiple failures during an operational phase;

– there is no possible maintenance action during an operational phase, and this can only be carried out at the beginning of the following day when powering the system;

– strictly speaking, a system consists of different elements as well as packaging, making it possible to connect them electronically and mechanically;

– for parallel systems, only active redundancy is considered. The same will be true for k-out-of-n redundant systems.

We will first explain the theoretical expressions of the reliability function and the MTTF for the three main families of systems and then present an example that introduces maintenance actions. The reliability function will then allow us to obtain the expression of the probability density by simple derivation and therefore that of the Rocof, whereas the MTTF will define the asymptotic value of the Rocof.

3.2. Maintenance-free systems

3.2.1. *Concept of the default rate*

Reliability is the study of the occurrence of failures over time. These failure events are random; that is, they cannot be known in advance, which significantly complicates everything. To model them, we use the notion of a random variable that will be denoted by T from here on. In this section, there are no *maintenance actions*, and the term "intrinsic reliability" can be employed for a component. We first examine the different types of failure. We can distinguish three main categories of failure, namely:

– Failures known as *"youth" failures* generally occur very early in the lifecycle of a product. Youth failures are usually due to manufacturing defects. They thus concern only a small part of the population.

– *"Catalectic" failures* occur because they are sudden, abrupt and independent of the time elapsed. This type of failure can therefore be observed at any time in the lifecycle of a product. They are generally due to accidental overloads (thermal, mechanical, electrical, etc.). They generally do not affect the entire product population, and the risk that they may occur can be reduced by robustness testing, derating rules, etc.

– *"Aging" failures* are observed throughout the entire product range in operation. Generally speaking, these failures are not observed during a product lifecycle, except for specific "limited-life" components, also known as premature aging components, due to incorrect sizing, batches of defective components, etc. These failures affect the entire population and, as such, must be delayed beyond the introduction stage of the product. For this purpose, design rules (derating rules, worst-case analysis, thermal, mechanical, electrical simulation, etc.) and specific aging tests can be implemented.

First, let us look at intrinsic reliability. By this, we refer to the reliability of a component, a board, a product, etc., without maintenance actions. To estimate it and, in particular, to know what type of failure has to be addressed, the most commonly used parameter is the (instantaneous) failure rate, denoted by h. This is defined by:

$$h(t) = lim_{\Delta t \to 0} \frac{P[t \leq T < t + \Delta t \; / T > t]}{\Delta t} \qquad [3.1]$$

Let us analyze this equation and the following conventions. The term P means "probability", and the symbol "/" means "knowing that". The limit "Lim" represents the instantaneous nature of the default rate. Therefore, the previous equation can be expressed as:

Probability that the product has a failure between "t and t+dt" knowing that the product was functional (no failure) at time "t".

The most commonly used mathematical object for modeling the failure rate, regardless of the type of failure (youth, accidental or aging), is the Weibull distribution. Under this assumption, it is defined by:

$$h(t) = \left(\frac{\beta}{\eta}\right).\left(\frac{t}{\eta}\right)^{\beta-1} \quad [3.2]$$

η is a scale parameter (usually time-related) and represents the characteristic lifetime, characterized by the fact that at time t = η, we obtain ~ 63.2% (1 – exp(-1)) of failure, regardless of the value of the parameter β and thus regardless of the type of failure. This modeling is interesting for the following reasons:

– The mathematical formulation is simple since it is an easily modifiable power function (continuous, differentiable, integrable, etc.).

– Depending on the parameter β, this function is decreasing ($\beta < 1$), constant ($\beta = 1$) or increasing ($\beta > 1$). In other words, it can represent the three types of failures previously defined.

– The parameter β has physical significance because it represents the aging kinetics of the observed failure mechanism. Indeed, we emphasize the random nature of the failure moments (the components under test are assumed to be all identical). This means that instead of having a single true value if failures are purely deterministic, we observe a dispersion of failure moments over time. In fact, the parameter β is the image of this dispersion, and the larger it is, the less dispersed the failure times. Ultimately, if β were infinite, all of the times of failure would be identical, which is never the case in reality. For further details, we refer the reader to Bayle (2019).

The expression of the reliability function can be easily obtained from the failure rate. It is then shown that (Appendix 6):

$$R(t) = \exp\left(-\int_0^t h(u).du\right) \quad [3.3]$$

Similarly, a probability distribution MTTF is given by:

$$MTTF = \int_0^{+\infty} R(t).dt \quad [3.4]$$

Exponential distribution of parameter λ

In this case, we have $h(t) = \lambda$ from which:

$$R(t) = \exp(-\lambda.t) \quad [3.5]$$

and

$$MTTF = \frac{1}{\lambda} \quad [3.6]$$

Weibull distribution of parameters η and β

From equation [3.2], we obtain:

$$R(t) = \exp\left(-\left(\frac{t}{\eta}\right)^{\beta}\right) \quad [3.7]$$

and

$$MTTF = \eta.\Gamma\left(1 + \frac{1}{\beta}\right) \quad [3.8]$$

3.2.2. *Maintenance-free serial system*

Consider a serial system of "s" elements that can be illustrated as follows:

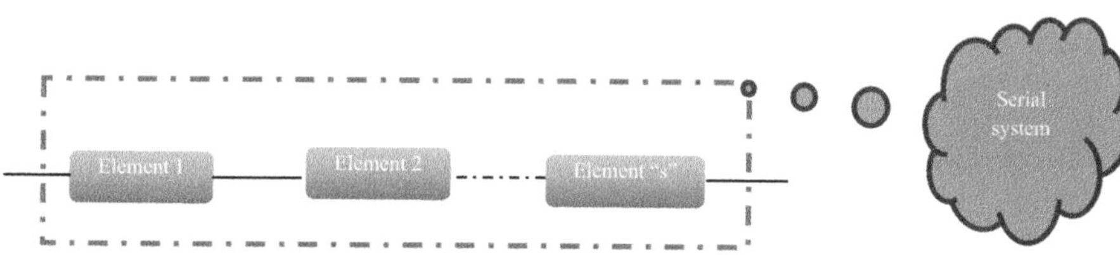

Figure 3.1. *Illustration of a serial system*

For a serial system to fail, only one of these elements has to fail. The system reliability function R_s is therefore written as:

$$R_S(t) = \prod_{i=1}^{s} R_i(t)$$

We express it in terms of the cumulative failure rate, that is:

$$R_S(t) = \prod_{i=1}^{s} exp\left(-\int_0^t h_i(u).du\right) = exp(-\sum_{i=1}^{s} H_i(t)) \quad [3.9]$$

The MTTF is therefore given by:

$$MTTF_S = \int_0^{+\infty} R_S(t).dt = \int_0^{+\infty} exp(-\sum_{i=1}^{s} H_i(t)).dt \quad [3.10]$$

Exponential distributions

We assume here that all of the elements follow an exponential distribution. This assumption is quite realistic, especially for electronic components (catalectic failures). The cumulative failure rate is then written as $H_i(t) = \lambda_i . t$. Therefore, from equation [3.9], the reliability function of the system is written as:

$$R_S(t) = exp\left(-\sum_{i=1}^{s} \lambda_i . t\right)$$

We can thus see that the reliability function of the system follows an exponential distribution of parameter $\sum_{i=1}^{s} \lambda_i$.

The corresponding MTTF is obtained from equation [3.10], that is:

$$MTTF_S = \int_0^{+\infty} exp(-\sum_{i=1}^{s} \lambda_i . t). dt = \frac{1}{\sum_{i=1}^{s} \lambda_i} \qquad [3.11]$$

Weibull distributions

Here, we assume that all of the elements follow a Weibull distribution, which is rarely the case in practice. The cumulative failure rate is then written as: $H_i(t) = \left(\frac{t}{\eta_i}\right)^{\beta_i}$. Hence, from equation [3.9], the reliability function of the system is written as:

$$R_S(t) = exp\left(-\sum_{i=1}^{s} \left(\frac{t}{\eta_i}\right)^{\beta_i}\right) \qquad [3.12]$$

We see that the reliability function no longer follows a Weibull distribution and that it cannot be rigorously modeled by a simple distribution.

In addition, the MTTF is obtained from equations [3.10] and [3.12] but does not have an explicit solution, as shown below:

$$MTTF_S = \int_0^{+\infty} exp\left(-\sum_{i=1}^{s} \left(\frac{t}{\eta_i}\right)^{\beta_i}\right). dt \qquad [3.13]$$

Exponential distributions and one Weibull distribution

In industry, problems in which exponential distributions have to be addressed often exist. However, in certain applications, there is sometimes an element (mechanical, electromechanical, optical, sensor, etc.) that follows a Weibull distribution. Let us therefore assume that the system has "s-1" elements following an exponential distribution and that the "*s*th" element follows a Weibull distribution. From equation [3.9], we can then write the following:

$$R_S(t) = exp\left(-\sum_{i=1}^{s-1} \lambda_i . t - \left(\frac{t}{\eta_s}\right)^{\beta_s}\right) \qquad [3.14]$$

REMARK.–

The choice of the value "s" is arbitrary but has no impact since the system is serial.

A simple distribution to model the reliability function of the system can no longer be found. The MTTF is obtained from equations [3.10] and [3.14], that is:

$$MTTF_S = \int_0^{+\infty} exp\left(-\sum_{i=1}^{s-1} \lambda_i . t - \left(\frac{t}{\eta_S}\right)^{\beta_S}\right) . dt \quad [3.15]$$

The previous equation shows that there is no simple analytic expression either.

EXAMPLE 3.1.– Element 1 exponential distribution of parameters λ = 0.001 and Element 2 Weibull distribution of parameters η = 500 and β = 3.7: we obtain the following reliability function:

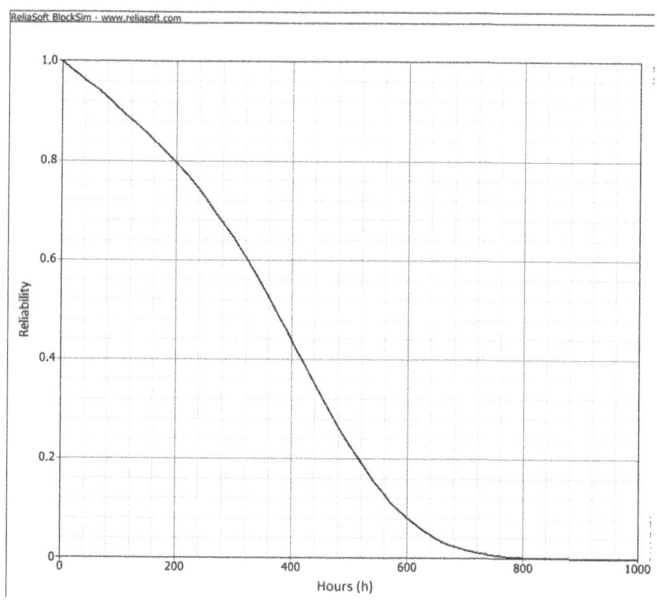

Figure 3.2. *Example of a maintenance-free serial system consisting of an exponential and a Weibull distribution*

3.2.3. *Maintenance-free parallel system*

A parallel system, composed of "p" identical elements, can be represented according to Figure 3.3.

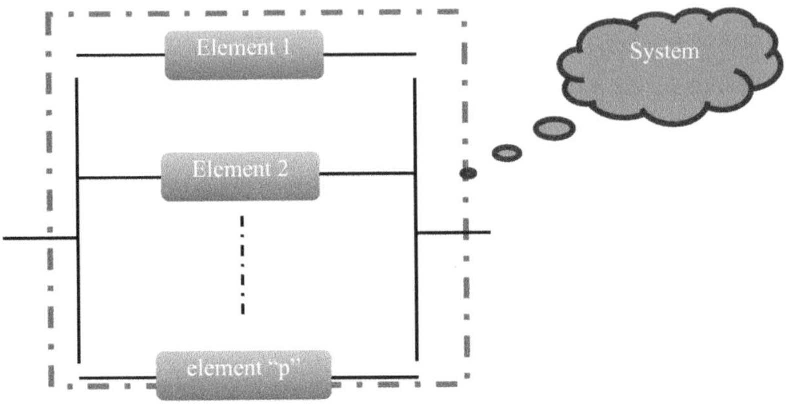

Figure 3.3. *Illustration of a parallel system*

For a parallel system to fail, all its elements must fail. The system probability of failure is therefore written as:

$$F_P(t) = \prod_{i=1}^{p} F_i(t)$$

Since all of the elements are identical, we obtain more simply:

$$F_P(t) = F(t)^p$$

Consequently, the reliability function is written as:

$$R_P(t) = 1 - F_P(t) = 1 - F(t)^p = 1 - \left(1 - R(t)\right)^p$$

From Newton's binomial formula, we can write the following:

$$R_P(t) = 1 - \sum_{j=0}^{p} C_p^j . (-1)^j . R(t)^j$$

or still

$$R_P(t) = \sum_{j=1}^{p} C_p^j . (-1)^{j+1} . R(t)^j$$

Using the relationship between the failure rate and reliability function, we obtain:

$$R_P(t) = \sum_{j=1}^{p} C_p^j.(-1)^{j+1}.exp\left(-j.\int_0^t h(u).du\right)$$

that is, finally:

$$R_P(t) = \sum_{j=1}^{p} C_p^j.(-1)^{j+1}.exp\big(-j.H(t)\big) \quad [3.16]$$

The corresponding MTTF is given by:

$$MTTF_P = \int_0^{+\infty} \sum_{j=1}^{p} C_p^j.(-1)^{j+1}.exp\big(-j.H(t)\big).dt$$

that is:

$$MTTF_P = \sum_{j=1}^{p} C_p^j.(-1)^{j+1}.\int_0^{+\infty} exp\big(-j.H(t)\big).dt \quad [3.17]$$

Exponential case

The cumulative failure rate is written as: $H(t) = \lambda.t$. From equation [3.16], we obtain:

$$R_P(t) = \sum_{j=1}^{p} C_p^j.(-1)^{j+1}.exp(-j.\lambda.t) \quad [3.18]$$

REMARK.–

Therefore, we do not obtain an exponential distribution for a parallel system, even if the distribution of each element parallelized is exponential.

The corresponding MTTF is obtained from equation [3.17] and Appendix 1, where $\mathrm{MTTF_e} = \frac{1}{\lambda}$.

$$MTTF_P = \mathrm{MTTF}_e.\sum_{j=1}^{p} \frac{1}{j} \quad [3.19]$$

Weibull case

The cumulative failure rate is written as:

$$H(t) = \left(\frac{t}{\eta}\right)^{\beta}$$

The reliability function is obtained from equation [3.16], that is:

$$R_S(t) = \sum_{j=1}^{p} C_p^j . (-1)^{j+1} . exp\left(-j . \left(\frac{t}{\eta}\right)^{\beta}\right) \qquad [3.20]$$

The MTTF is obtained from equation [3.17], namely:

$$MTTF_P = \sum_{j=1}^{p} C_p^j . (-1)^{j+1} . \int_0^{+\infty} exp\left(-j . \left(\frac{t}{\eta}\right)^{\beta}\right) . dt$$

Let $u = j . \left(\frac{t}{\eta}\right)^{\beta}$ or still $u = \left(\frac{t}{\eta . j^{-1/\beta}}\right)^{\beta}$.

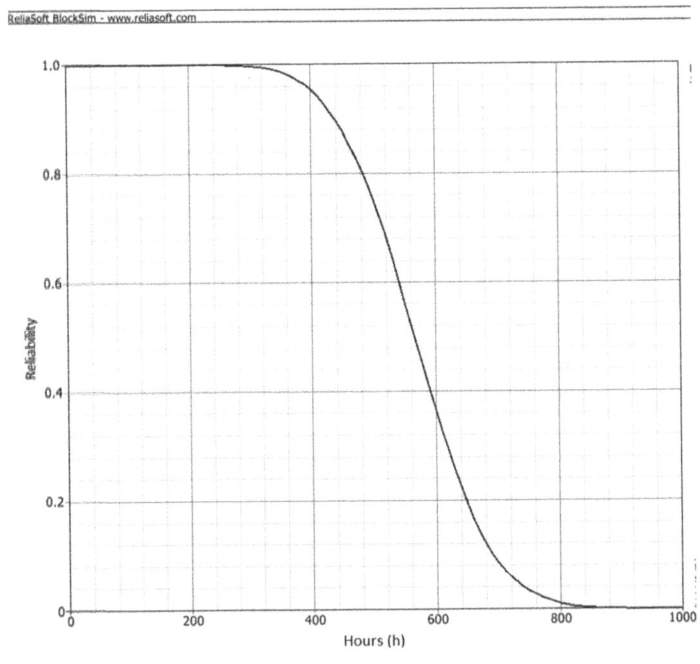

Figure 3.4. *Reliability function for a parallel system of three identical maintenance-free Weibulls*

We thus have $du = \beta . \left(\frac{t}{\eta . j^{-1/\beta}}\right)^{\beta-1} . \left(\frac{dt}{\left(\eta . j^{-1/\beta}\right)^{\beta-1}}\right)$ from which:

$$MTTF_P = \sum_{j=1}^{p} C_p^j.(-1)^{j+1}.\int_0^{+\infty} \frac{exp(-u).\left(\eta.j^{-1/\beta}\right)^{\beta-1}}{.\left(\frac{t}{\eta.j^{-1/\beta}}\right)^{\beta-1}}.du$$

That is, finally, by denoting $\text{MTTF}_W = \eta.\Gamma\left(1+\frac{1}{\beta}\right)$.

$$MTTF_P = \text{MTTF}_W.\sum_{j=1}^{p} \frac{C_n^j.(-1)^{j+1}}{j^{1/\beta}} \quad [3.21]$$

EXAMPLE 3.2.– Three Weibull distribution elements of parameters η = 500 and β = 3.7.

3.2.4. *Maintenance-free "k-out-of-n" redundancy*

Here, we address "active" redundancy, that is, power is applied to the elements. For the system to fail, there must be at least "k+1" faulty elements. This is illustrated by Figure 3.5.

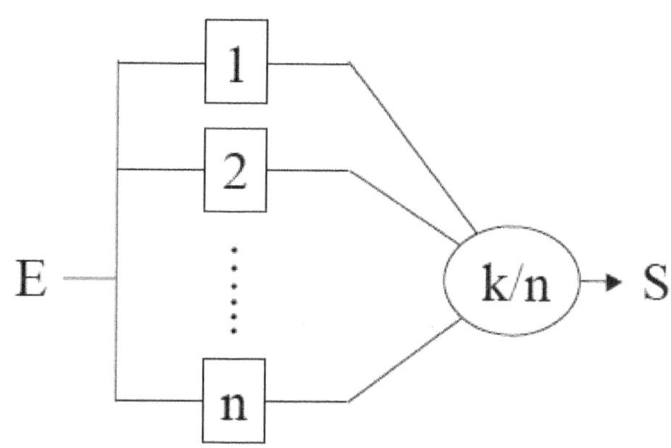

Figure 3.5. *Illustration of k-out-of-n active redundancy*

The reliability function of such a system is written as (Rausand and Hoyland 2004):

$$R_{kn}(t) = \sum_{j=k}^{n} C_n^j.R(t)^j.(1-R(t))^{n-j} \quad [3.22]$$

That is, with the Newton binomial formula:

$$R_{kn}(t) = \sum_{j=k}^{n} C_n^j . R(t)^j . \sum_{i=0}^{n-j} C_{n-j}^i . (-1)^i . R(t)^i$$

Namely:

$$R_{kn}(t) = \sum_{j=k}^{n} \sum_{i=0}^{n-j} C_n^j . C_{n-j}^i . (-1)^i . R(t)^{i+j}$$

which is:

$$R_{kn}(t) = \sum_{j=k}^{n} \sum_{i=0}^{n-j} \left(C_n^j . C_{n-j}^i . (-1)^i . exp\big(-(i+j). H(t)\big)\right) \quad [3.23]$$

The MTTF is therefore given by:

$$MTTF_{kn} = \sum_{j=k}^{n} \sum_{i=0}^{n-j} C_n^j . C_{n-j}^i . (-1)^i . \int_0^{+\infty} exp\big(-(i+j). H(t)\big). dt \quad [3.24]$$

Exponential case

From equation [3.23], the reliability function is written as:

$$R_{kn}(t) = \sum_{j=k}^{n} \sum_{i=0}^{n-j} C_n^j . C_{n-j}^i . (-1)^i . exp(-\lambda . t)^{i+j}$$

That is, finally:

$$R_{kn}(t) = \sum_{j=k}^{n} \sum_{i=0}^{n-j} C_n^j . C_{n-j}^i . (-1)^{i+1} . exp(-(i+j). \lambda . t) \quad [3.25]$$

The MTTF is given, according to Appendix 1 and by setting $MTTF_e = \frac{1}{\lambda}$, by:

$$MTTF_{kn} = MTTF_e . \sum_{i=k}^{n} \frac{1}{i} \quad [3.26]$$

Weibull case

The reliability function is obtained from equation [3.23], that is:

$$R_{kn}(t) = \sum_{j=k}^{n} \sum_{i=0}^{n-j} C_n^j . C_{n-j}^i . (-1)^i . exp\left(-(i+j). \left(\frac{t}{\eta}\right)^{\beta}\right)$$

The MTTF is obtained from equation [3.24], namely:

$$MTTF_{kn} = \sum_{j=k}^{n}\sum_{i=0}^{n-j} C_n^j . C_{n-j}^i . (-1)^i . \int_0^{+\infty} exp\left(-(i+j).\left(\frac{t}{\eta}\right)^{\beta}\right).dt$$

or still:

$$MTTF_{kn} = \text{MTTF}_W . \sum_{j=k}^{n}\sum_{i=0}^{n-j} \frac{C_n^j . C_{n-j}^i . (-1)^i}{(i+j)^{1/\beta}} \quad [3.27]$$

EXAMPLE 3.3.– k = 3 out of n = 10 Weibull of parameters $\eta = 500$ and $\beta = 3.7$.

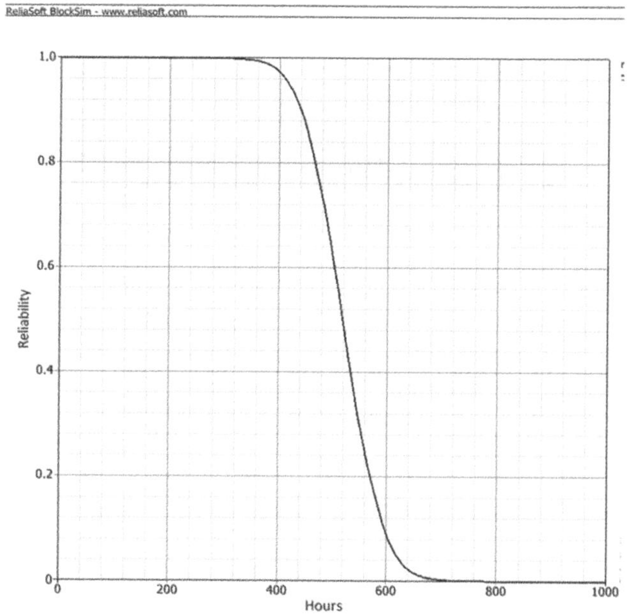

Figure 3.6. *Example of a reliability function of a k-out-of-n redundancy*

3.3. Maintenance-free systems

In Chapter 2, we described (for maintenance-free systems) how to calculate the reliability function and the MTTF considering different types of systems. Nonetheless, for most applications, the systems are maintained under operational conditions and therefore undergo maintenance actions. Therefore, as soon as the first failure occurs, the failure rate can no longer be used.

REMARK.–

Under operational conditions, it is not the "intrinsic reliability" of a "single component" that is the quantity of interest. This is the reason why, in predictive reliability methods, reliability prediction concerns that of the component and its implementation on the printed circuit board.

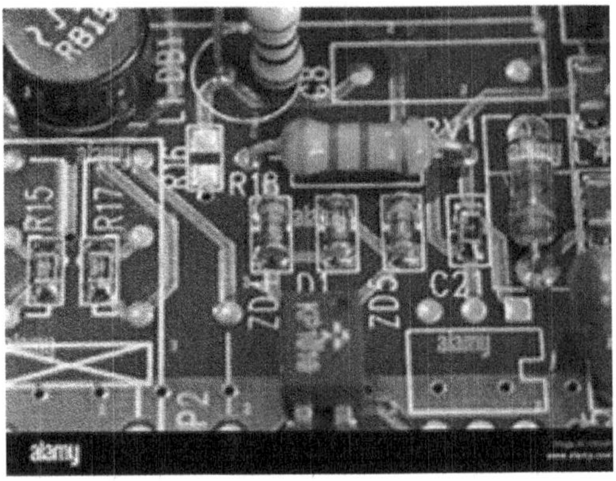

Figure 3.7. *Example of electronic components mounted on printed circuit*

The sought reliability is that of the function performed by the component and its transfer, that is, in the event of failure in the component, it is that of all of the components (admittedly identical) that have performed the desired function, particularly their number and the times at which they failed.

3.3.1. *Concept of failure intensity and occurrence rate of Rocof failures*

We observed in section 3.2.1 that the failure rate only makes sense for maintenance-free applications. What if maintenance is indeed enabled? The notion of failure intensity was then proposed.

In accordance with the previous remark, it was defined by:

$$\lambda(t) = lim_{\Delta t \to 0} \frac{P[N(t,t+\Delta t)/H_t]}{\Delta t} \quad [3.28]$$

where H_t represents the history before time "t", namely, the number of failures observed as well as the number of times when they occurred. N(t, t+Δt) represents the number of failures in (t, t+Δt).

Generally speaking, the failure intensity is random (while the failure rate is deterministic) since failure times are deterministic.

These two notions can be illustrated in Figure 3.8.

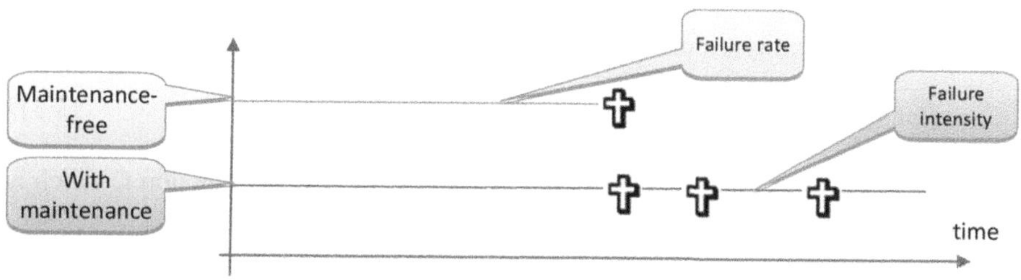

Figure 3.8. *Illustration of the confusion for maintenance-enabled and maintenance-free systems*

This explains the amalgamation that is often made for catalectic failures for maintenance-enabled and maintenance-free systems.

If we look into the evolution of the number of failures over time, we obtain the following illustration.

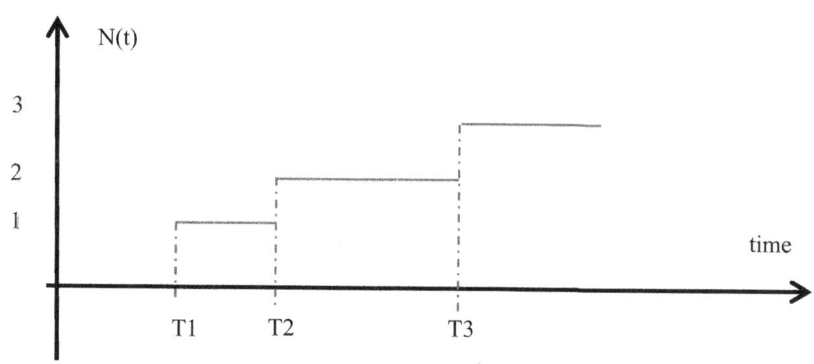

Figure 3.9. *Illustration of the failure counting process*

Since the failure times are random, the failure counting process is therefore also random. To obtain a deterministic variable, we look at its average value defined by:

$$m(t) = E[N(t)] \tag{3.29}$$

where E represents the mathematical expectation.

The "Rocof" failure occurrence rate is then defined by:

$$Rocof(t) = \frac{\mathrm{d}}{\mathrm{dt}} m(t) = \frac{\mathrm{d}}{\mathrm{dt}} E[N(t)] \quad [3.30]$$

In the case of *perfect maintenance* (instantaneous replacement of the defective entity with a new entity), it can be shown that (Basu and Rigdon 2000):

$$Rocof(t) = \sum_{i=1}^{+\infty} f^{<i>}(t) \quad [3.31]$$

where $f^{<i>}(t)$ is the convolutional product "i times" by itself of function f, which is the probability density of the underlying failure mechanism.

For the exponential distribution of parameter λ, the density is given by $f(t) = \lambda.\exp(-\lambda.t)$.

The Rocof is then given by (Bayle 2019):

$$Rocof(t) = \lambda \quad [3.32]$$

For the normal distribution of parameters μ and σ, the corresponding Rocof is given by (Bayle 2019):

$$Rocof(t) = \sum_{i=1}^{+\infty} \varphi(t, i.\mu, \sqrt{i}.\sigma) \quad [3.33]$$

REMARK.–

The normal distribution has an explicit solution for the calculation of the Rocof; although it is not very suitable as a reliability model (because of its symmetry with respect to the average and its domain of definition, including negative values), it can be used to approximate, under certain conditions, the Rocof of a Weibull distribution that has no explicit solution.

The asymptotic value of Rocof is given by (Basu and Rigdon 2000):

$$\mathrm{Rocof}_\infty = \lim_{t \to +\infty} \mathrm{Rocof(t)} = \frac{1}{\mathrm{MTTF}} \quad [3.34]$$

3.3.2. *Serial system with maintenance*

3.3.2.1. *Element-level renewal*

We previously reported that a serial system is faulty when one of its elements is faulty. When the failed element is instantly replaced with a new element in a series system, we address an overlay of simple renewal processes. For example, assume that element "j" among the "s" elements of the system fails, and then that element "s" is defective. We obtain the following illustration.

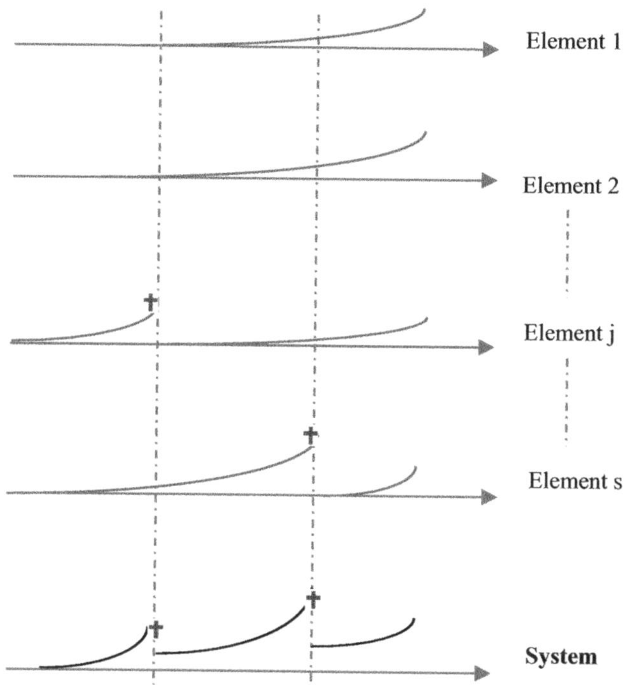

Figure 3.10. *Illustration of failures for a serial system with element renewal*

It can thus be written that, for an overlapping renewal process (Appendix 2):

$$Rocof_S(t) = \sum_{i=1}^{s} Rocof_i(t) \qquad [3.35]$$

Generally, we do not obtain a system-level renewal process unless all of the elements follow an exponential distribution. On the other hand, at the element level, the asymptotic value of the Rocof is equal to the inverse of the MTTF of the distribution under consideration. Additionally, the asymptotic value of the Rocof of the serial system is given by:

$$lim_{t\to+\infty} Rocof_S(t) = lim_{t\to+\infty} \sum_{i=1}^{s} Rocof_i(t) = \sum_{i=1}^{s} lim_{t\to+\infty} Rocof_i(t)$$

That is:

$$Rocof_{\infty} = lim_{t\to+\infty} Rocof_S(t) = \sum_{i=1}^{S} \frac{1}{MTTF_i} \quad [3.36]$$

Exponential distribution of parameter λ_i (catalectic failures)

Since the Rocof of an exponential parameter distribution λ is equal to λ (Bayle 2019), the system Rocof is given by:

$$Rocof_S(t) = \sum_{i=1}^{S} \lambda_i \quad [3.37]$$

According to equations [3.34], the asymptotic value of the Rocof for a serial system can be written as:

$$Rocof_{\infty} = \frac{1}{MTTF_S} = \sum_{i=1}^{S} \lambda_i = Rocof_S(t) \quad [3.38]$$

REMARK.–

A comparison of equations [3.3] for maintenance-free systems and [3.28] for maintenance-enabled systems reveals that they are identical and yield the same numerical results. However, it is clear that the theoretical principles are completely different. On the one hand, the failure rate is used and, on the other hand, the rate of failure occurrence or Rocof.

Normal distribution example (aging failures)

For aging failures, we use a normal approximation (Bayle 2019) to obtain a theoretical formulation of the number of failures:

$$Rocof(t) = \sum_{k=1}^{+\infty} \varphi(t, k.\mu, k.\sqrt{\sigma})$$

where φ is the probability density of the normal distribution.

Therefore, the system Rocof is given by:

$$Rocof_S(t) = \sum_{i=1}^{S} Rocof_i(t) = \sum_{i=1}^{S} \sum_{j=1}^{+\infty} \varphi(t, j.\mu_i, \sqrt{j}.\sigma_i) \quad [3.39]$$

According to equation [3.36], its asymptotic value is equal to: $Rocof_{\infty} = \frac{1}{\sum_{i=1}^{n} \mu_i}$.

EXAMPLE 3.4.– Consider the example of a system with two elements in series with the following characteristics: Element 1: Normal(500; 100) Element 1: Normal(1,000; 150).

The Rocof of the two normal distributions is shown in Figure 3.11.

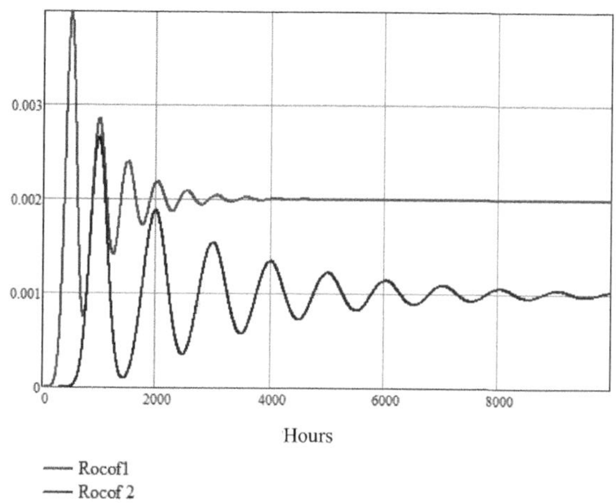

Figure 3.11. *Rocof of two normal distributions with maintenance at element level*

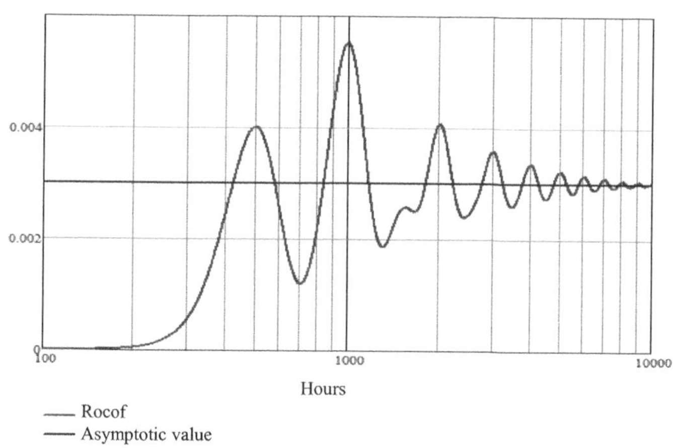

Figure 3.12. *Rocof of two normal distributions in series with maintenance at element level*

The Rocof of the serial system is then the sum of the two Rocofs.

According to equation [3.36], the asymptotic value of the Rocof is: $Rocof_\infty = \frac{1}{\mu 1} + \frac{1}{\mu 2} = 0.003$.

Simulations

We have the flowchart shown in Figure 3.13.

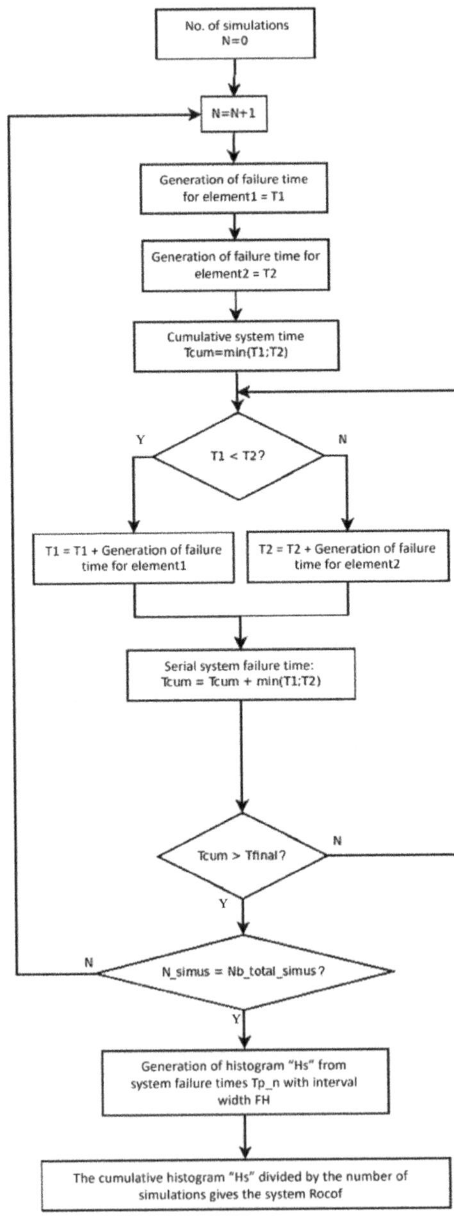

Figure 3.13. *Component renewal simulation flow chart for a serial system*

We obtain the following figure:

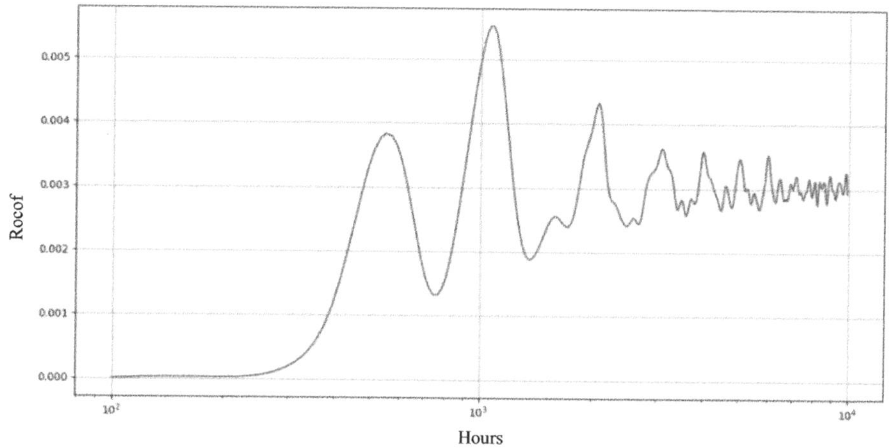

Figure 3.14. *Simulation of the Rocof of two normal distributions in series with maintenance at element level*

Weibull distributions of parameters η_i and β_i

The Rocof is obtained from equation [3.31] but has no explicit solutions. A numerical resolution is therefore necessary.

In practice, the system can often be considered a serial system comprising an exponential element model of the electronic part and a Weibull element model of a subset with a limited lifecycle (electromechanical, mechanical, sensor, etc.).

The Rocof asymptotic value is obtained from equation [3.26] and is equal to:

$$Rocof_{\infty} = \sum_{i=1}^{s} \left(\frac{1}{\eta_i . \Gamma\left(1 + \frac{1}{\beta_i}\right)} \right)$$

EXAMPLE 3.5.– Exponential (0.001) (actually Weibull with β = 1) and Weibull (1,500; 5).

The corresponding Rocof is shown in Figure 3.15.

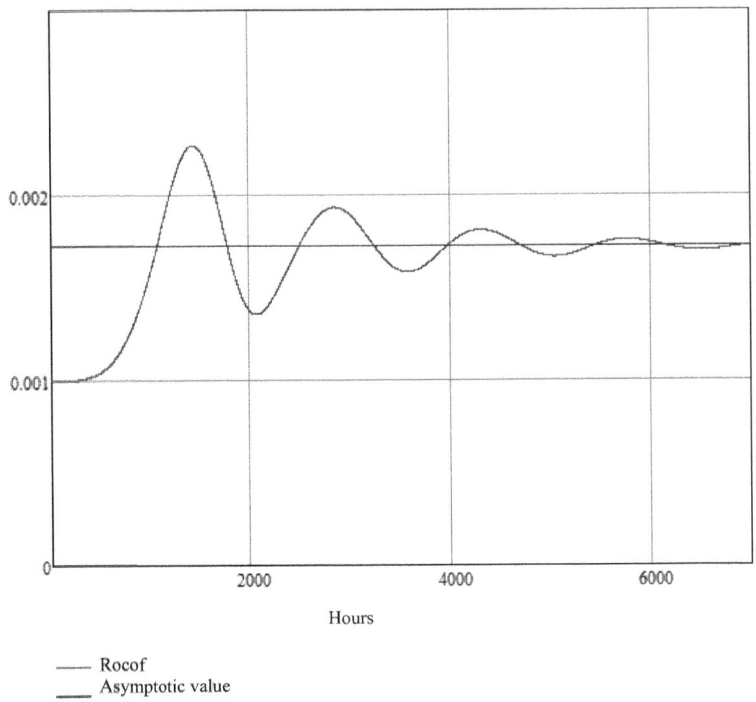

Figure 3.15. *Rocof of two Weibull distributions in series with maintenance at element level*

The asymptotic value of the Rocof is:

$$Rocof_{\infty} = \frac{1}{MTTF_{exp}} + \frac{1}{MTTF_{Wei}} = \frac{1}{1/\lambda} + \frac{1}{\eta.\Gamma\left(1+\frac{1}{\beta}\right)}$$

that is:

$$Rocof_{\infty} = \lambda + \frac{1}{\eta.\Gamma\left(1+\frac{1}{\beta}\right)}$$

Simulation

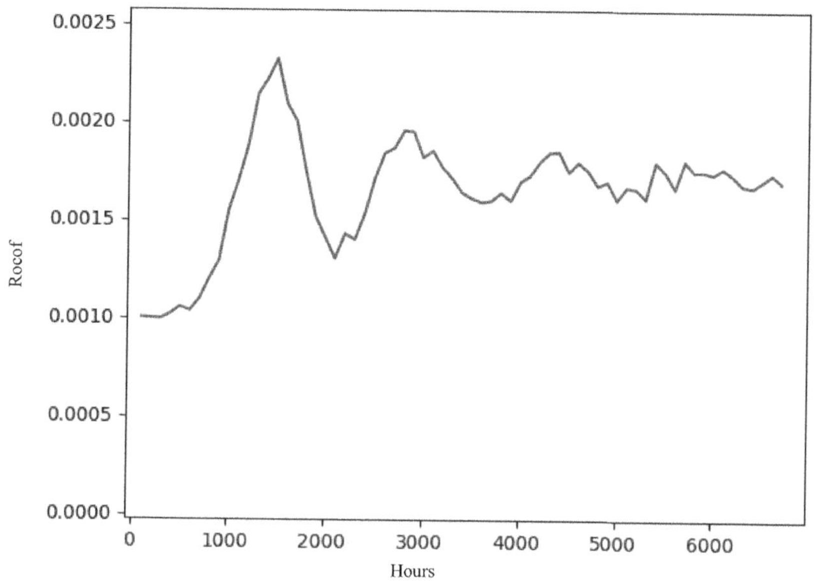

Figure 3.16. *Rocof of two Weibull distributions in series with maintenance at element level*

3.3.2.2. *System-level renewal*

As soon as one element fails, the system fails, and with this strategy, instead of replacing the failed element with a new one as previously described, all of the elements of the system are replaced.

The system Rocof is no longer equal to the sum of the Rocof of each of the elements since these are renewed even if they have not failed. However, because all of the elements are replaced, this is viewed as a system-level renewal process. To compute the system Rocof, it is therefore necessary to compute, in accordance with equation [3.31], the probability density of the serial system (Appendix 3), that is:

$$f_S(t) = \sum_{i=1}^{s} \left(f_i(t). \prod_{\substack{j=1 \\ j \neq i}}^{s} R_j(t) \right)$$

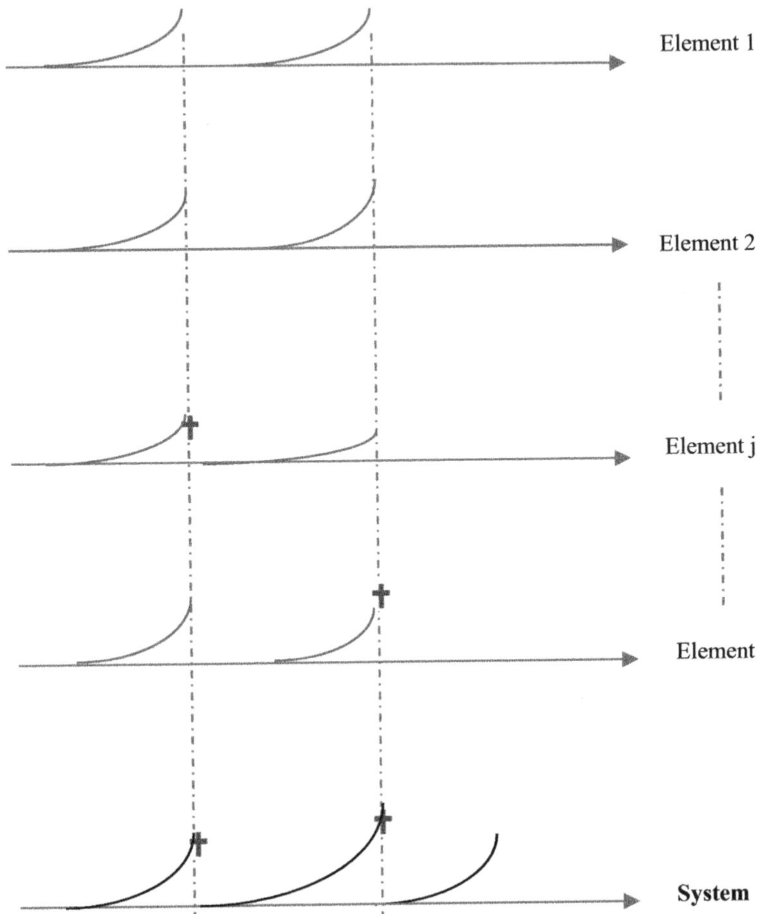

Figure 3.17. *Illustration of the failures for a serial system with system renewal*

REMARK.–

Except for the exponential distribution, the density of a series system does not follow a known distribution.

The Rocof of the system is therefore equal to:

$$Rocof_S(t) = \sum_{i=1}^{+\infty}\left(\sum_{j=1}^{S}\left(f_i(t).\prod_{\substack{j=1\\ j\neq i}}^{S} R_j(t)\right)\right)^{<i>} \qquad [3.40]$$

Exponential distribution

In this case, the system probability density follows an exponential distribution of parameter $\sum_{i=1}^{S}\lambda_i$.

Consequently, we obtain the same equation as equation [3.37]. On the other hand, except for an exponential distribution, the Rocof for a serial system with system-level renewal has no explicit solutions.

EXAMPLE 3.6.– As an illustration, consider the example of a system with two elements and the following characteristics: Element 1: Normal(500; 100); Element 2: Normal(1,000; 150).

In the case of the normal distribution, the expression of the MTTF from equation [3.10] is not directly exploitable.

It is therefore preferable in this particular case to reason in the following way:

We have $R_S(t) = R_1(t).R_2(t) = \frac{1}{2}.\left(1 - erf\left(\frac{t-\mu_1}{\sigma_1.\sqrt{2}}\right)\right).\frac{1}{2}.\left(1 - erf\left(\frac{t-\mu_2}{\sigma_2.\sqrt{2}}\right)\right)$, which is

$R_S(t) = \frac{1}{4}.\left(1 - erf\left(\frac{t-\mu_1}{\sigma_1.\sqrt{2}}\right)\right).\left(1 - erf\left(\frac{t-\mu_2}{\sigma_2.\sqrt{2}}\right)\right)$, where erf is the error function.

Consequently, the MTTF is equal to:

$$MTTF_S = \int_0^{+\infty} \frac{1}{4}.\left(1 - erfc\left(\frac{t-\mu_1}{\sigma_1.\sqrt{2}}\right)\right).\left(1 - erf\left(\frac{t-\mu_2}{\sigma_2.\sqrt{2}}\right)\right).\mathrm{dt}$$

or still

$$MTTF_S = \frac{1}{4}.\int_0^{+\infty} erfc\left(\frac{t-\mu_1}{\sigma_1.\sqrt{2}}\right).erfc\left(\frac{t-\mu_2}{\sigma_2.\sqrt{2}}\right).\mathrm{dt}$$

The asymptotic value of the Rocof is thus equal to:

$$Rocof_{\infty} = \frac{4}{\int_0^{+\infty} erfc\left(\frac{t-\mu_1}{\sigma_1.\sqrt{2}}\right).erfc\left(\frac{t-\mu_2}{\sigma_2.\sqrt{2}}\right).dt}$$

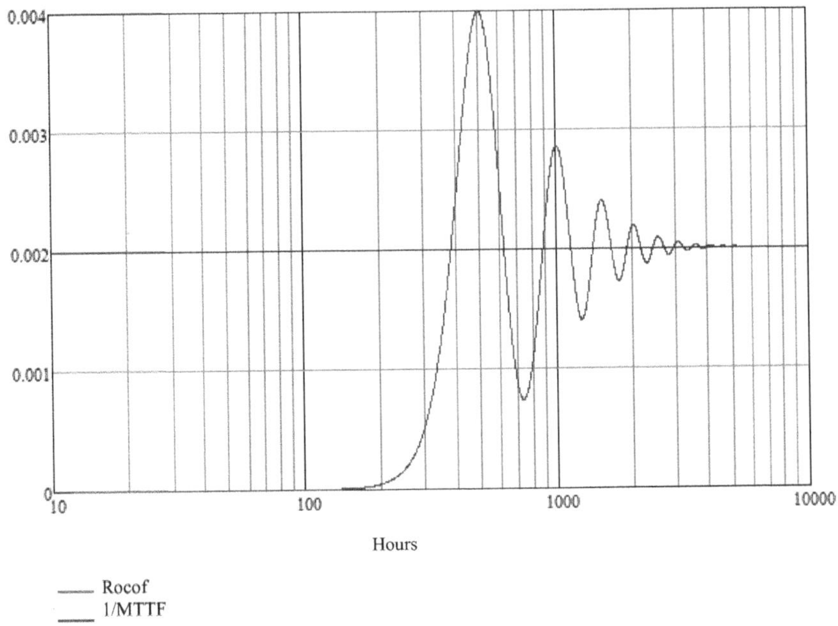

Figure 3.18. *Rocof of two normal distributions in series with maintenance at system level*

Simulation

Figure 3.19 shows the computational algorithm for a two-element serial system.

For more elements, the steps "Generation time at element failure" simply have to be repeated as many times as there are elements in series, and the system failure time Tp_n per min(T1, T2..., Tn) is calculated.

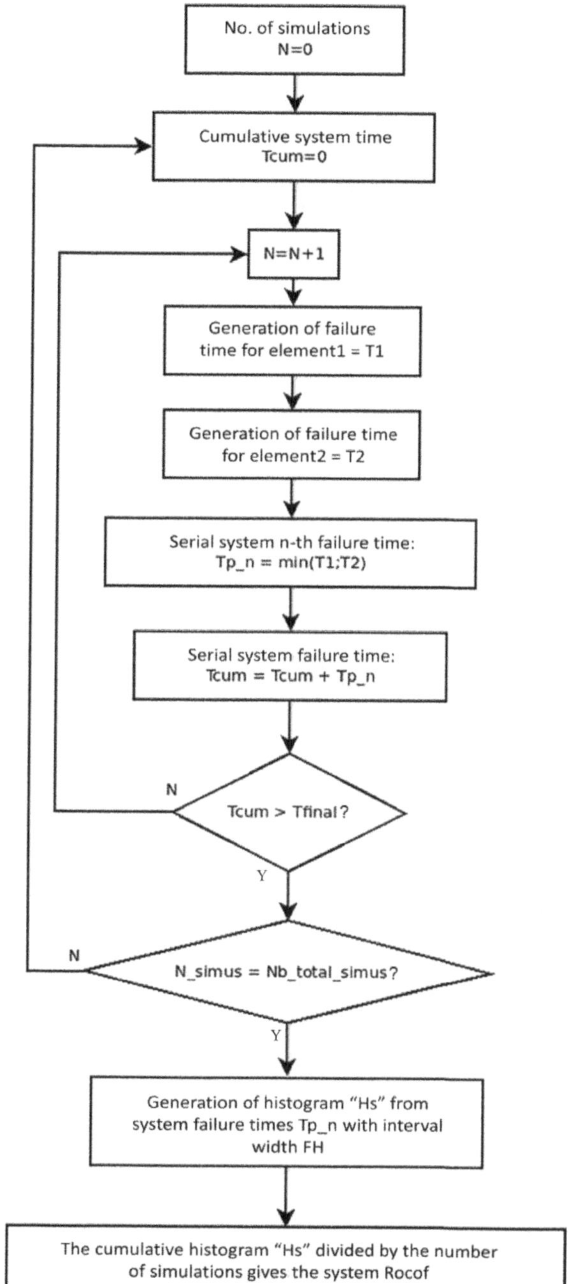

Figure 3.19. *Simulation flowchart for a two-element maintenance-enabled serial system*

We obtain the following Rocof:

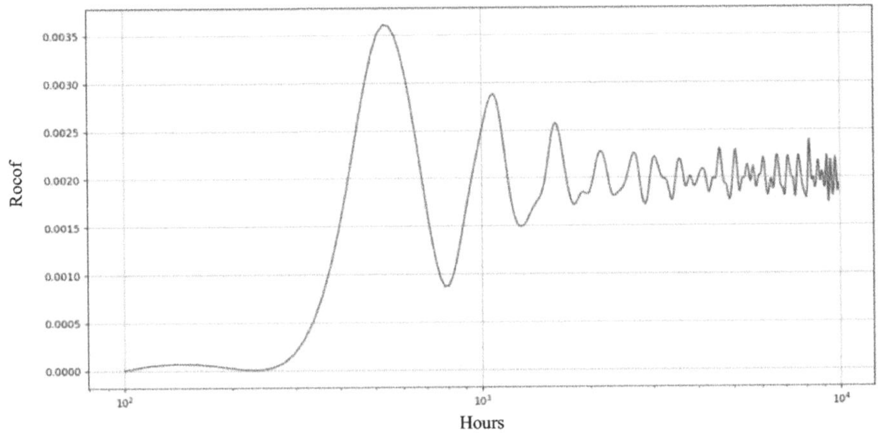

Figure 3.20. *Simulation of two normal distributions in series with maintenance at system level*

The curves obtained, as well as the asymptotic values, are in accordance with the previous numerical computation. Only simulation noise remains that could be reduced by increasing the number of simulations, which would also result in an increase in simulation time.

3.3.2.3. *Summary*

– *Element-level renewal*:

- The Rocof of the system is equal to the sum of the Rocof of each element.

- If the "n" elements of the system follow an exponential distribution of parameter λ_i, then the system Rocof also follows an exponential distribution of parameter $\sum_{i=1}^{n} \lambda_i$.

- As soon as one of the elements follows a Weibull distribution, the system Rocof has no explicit solutions.

- The asymptotic value of the system Rocof is equal to the sum of the inverses of the MTTFs of each element.

– *System-level renewal*:

- The Rocof is obtained from the probability density of the system. It has no explicit solutions unless all of the elements follow an exponential distribution. In this case, it is constant and equal to $\sum_{i=1}^{n} \lambda_i$.

- Its asymptotic value is equal to the inverse of the system MTTF and has no explicit solutions unless all of the elements follow an exponential distribution. In this case, it is equal to $\sum_{i=1}^{n} \lambda_i$.

3.3.3. *Parallel system with maintenance*

Unlike the serial system, even if the system is functional, we do not know which elements are failing. On the other hand, the type of renewal (element or system) has no impact because with a faulty system, with maintenance, all of the elements also fail. Let us consider a parallel system composed of "p" identical but independent elements.

3.3.3.1. *System-level renewal*

We obtain the illustration shown in Figure 3.21.

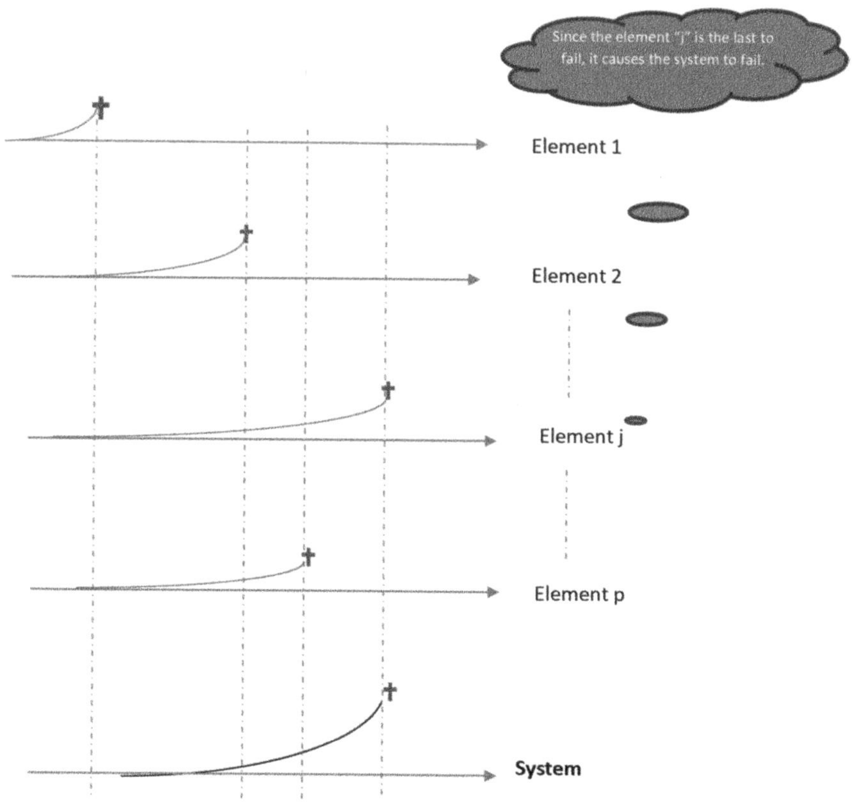

Figure 3.21. *Illustration of failures in a parallel system with element-level renewal*

The equivalent probability density is given by:

$$f_P(t) = \frac{dF(t)^p}{dt} = p.F(t)^{p-1}.f(t)$$

The Rocof is thus given by:

$$Rocof_P(t) = \sum_{j=1}^{+\infty} f_P(t)^{<j>} = \sum_{j=1}^{+\infty} \left(p.F(t)^{p-1}.f(t)\right)^{<j>} \quad [3.41]$$

Independent of the underlying probability distribution, the Rocof has no explicit solutions. Numerical computation or Monte Carlo simulations are therefore necessary. The asymptotic value of the Rocof for a parallel system is obtained from equations [3.5] and [3.34], that is:

$$Rocof_\infty = \frac{1}{\sum_{j=1}^{p} C_p^j.(-1)^{j+1}.\int_0^{+\infty} exp(-j.H(t)).dt} \quad [3.42]$$

Exponential distribution

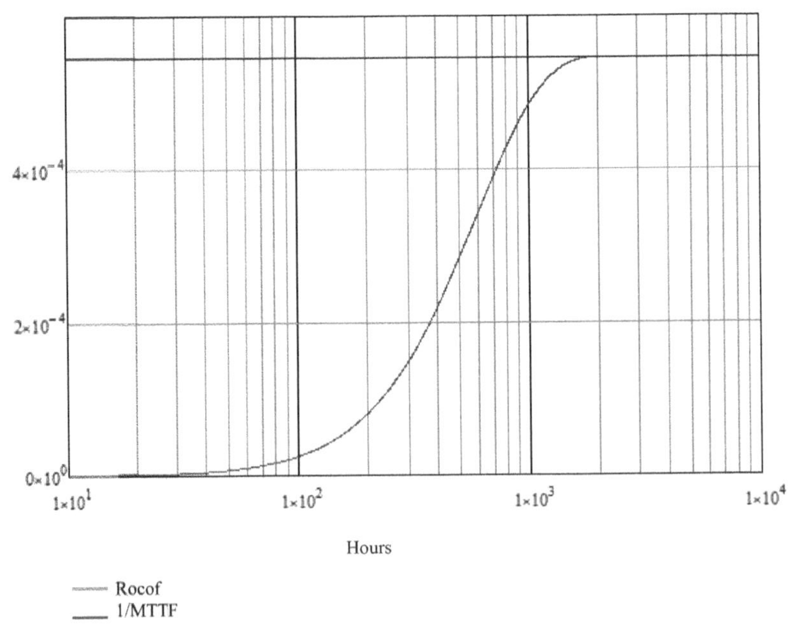

Figure 3.22. *Rocof of exponential distributions in parallel with maintenance at system level*

We have here $H(t) = \lambda.t$, from which:

$$Rocof_P(t) = p.\lambda.\sum_{j=1}^{+\infty}\left(\left(1-\exp(-\lambda.t)\right)^{p-1}.\exp(-\lambda.t)\right)^{<j>} \quad [3.43]$$

Unlike the serial system, the Rocof of a parallel system is no longer only a function of the scale parameter λ of the exponential distribution and does not have explicit solutions.

From equations [3.19] and [3.34], we have the expression of the asymptotic value of the Rocof, namely:

$$Rocof_\infty = \frac{1}{\mathrm{MTTF}_e.\sum_{j=1}^{p}\frac{1}{j}} \quad [3.44]$$

EXAMPLE 3.7.– Exponential (0.001) n = 3. We obtain the results shown in Figure 3.22.

Simulation

Figure 3.23 shows the calculation algorithm for a parallel system of three elements. For more elements, we simply have to repeat the steps "Generation element failure time" as many times as there are elements in series and compute the system failure time Tp_n by max (T1, T2, T3..., Tn). The principle is to determine the failure times (TTFs) of the parallel system and then to determine the frequency of these failure times by classes of width FH. The accumulation of FHs will make it possible to obtain the Rocof.

For example, if 10 simulations have obtained the following five failure times: {55, 265, 540, 560, 880}, then the frequency table will be given as:

FH	100	200	300	400	500	600	700	800	900
TTF	1	0	1	0	0	2	0	0	1

The cumulative table is thus as follows:

FH	100	200	300	400	500	600	700	800	900
TTF	1	1	2	2	2	4	4	4	5

By dividing by the number of simulations, the system Rocof is obtained:

FH	100	200	300	400	500	600	700	800	900
TTF	0.1	0.1	0.2	0.2	0.2	0.4	0.4	0.4	0.5

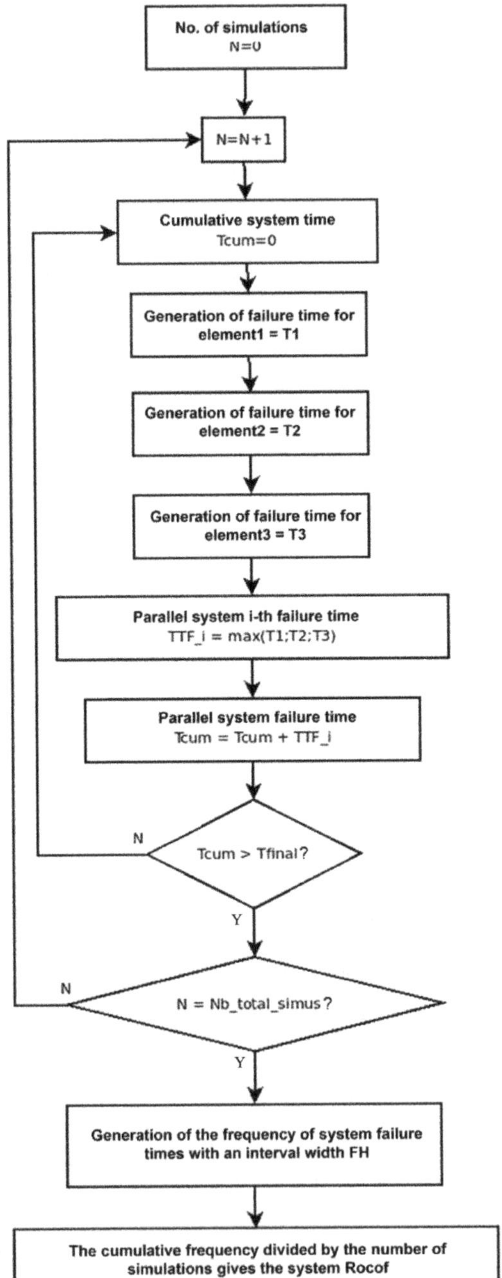

Figure 3.23. *Computational algorithm for the simulation of a three-element parallel system with maintenance*

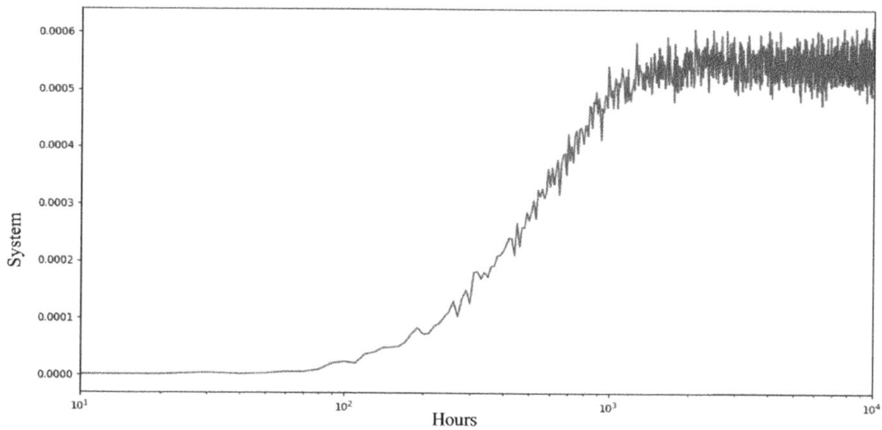

Figure 3.24. *Simulation of Rocof of exponential distributions in series with maintenance at system level*

The curves obtained, as well as the asymptotic values, are in accordance with the previous numerical computation. The only thing that remains is simulation noise that could be reduced by increasing the number of simulations, which would also result in an increase in simulation time.

Weibull distribution

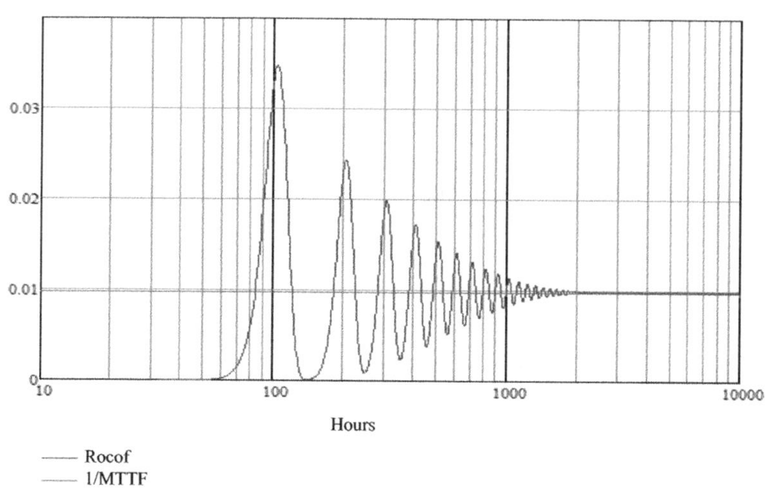

Figure 3.25. *Rocof of Weibull distributions in parallel with maintenance at system level*

Equation [3.41] shows that the Rocof does not have explicit solutions. The Rocof asymptotic value is given by equations [3.22] and [3.34], that is:

$$Rocof_{\infty} = \frac{1}{\mathrm{MTTF}_W \cdot \sum_{j=1}^{p} \frac{C_p^j \cdot (-1)^{j+1}}{j^{1/\beta}}} \quad [3.45]$$

EXAMPLE 3.8.– Weibull(100; 7) n = 2.

Simulation

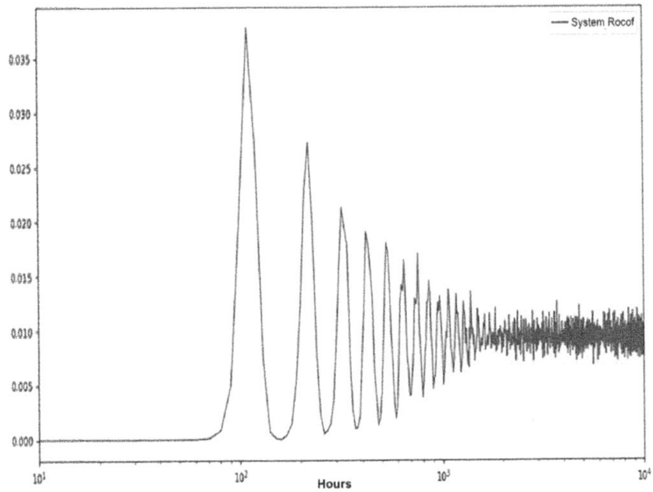

Figure 3.26. *Simulation of Weibull distributions in parallel with maintenance at system level*

The curve obtained as well as the asymptotic value are in agreement with the previous numerical computation.

3.3.3.2. *Item-level renewal*

The system will be defective when all of the elements are also defective. This brings us back to the problem of system-level renewal.

3.3.3.2.1. Summary

– *Element-level renewal*:

- Element-level renewal is identical to system renewal since all of the elements must be defective for the system to be defective.

– *System-level renewal*:

- The Rocof is obtained from the system probability density. It has no explicit solutions even if all of the elements follow an exponential distribution.

- For an exponential distribution, the asymptotic value is equal to $\frac{1}{\text{MTTF}_e \cdot \sum_{j=1}^{p} \frac{1}{j}}$, where MTTF is the MTTF of an exponential distribution.

- For a Weibull distribution, the asymptotic value is equal to $\frac{1}{\text{MTTF}_W \cdot \sum_{j=1}^{p} \frac{C_p^j \cdot (-1)^{j+1}}{j^{1/\beta}}}$, where MTTF_W is the MTTF of a Weibull distribution.

3.3.4. *k-out-of-n redundancy system with maintenance*

The reasoning is identical to that of the parallel system except that, instead of requiring "n" defective elements, only "k+1" are needed (k ≤ n) to have a system failure instead of "n", which changes the expression of the probability density for the computation of the Rocof or the reliability function for the MTTF.

3.3.4.1. *Element-level renewal*

We do not know any theoretical solutions for this maintenance strategy because we do not know which elements are faulty when the system fails.

Exponential distribution λ = 0.001; k = 2 and n = 4

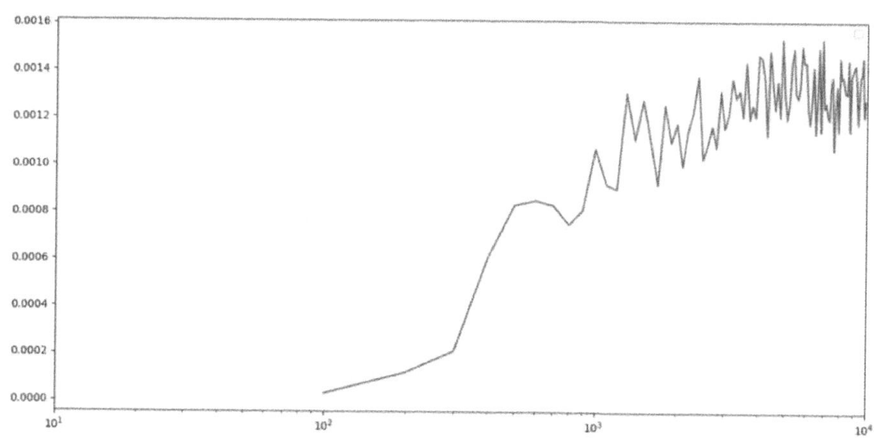

Figure 3.27. *Simulation of Rocof of exponential distributions in k/n redundancy with maintenance at element level*

The simulation yields the result shown in Figure 3.27.

Weibull: η = 100; β = 2; k = 2 and n = 4

The simulation yields the result shown in Figure 3.28.

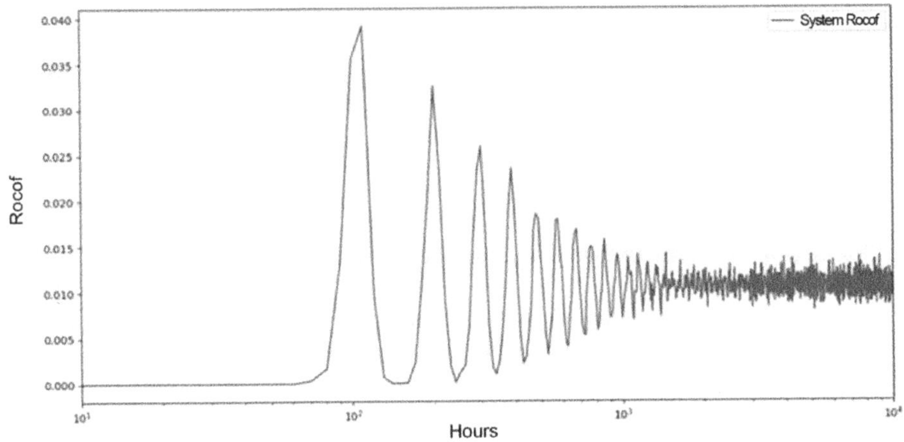

Figure 3.28. *Simulation of Rocof of Weibull distributions in k/n redundancy with maintenance at element level*

3.3.4.2. *System-level renewal*

Since we do not have any information about failure at the element level, the system will fail when "k+1" elements fail as well.

The equivalent probability density is given by:

$$f_S(t) = -\frac{dR_S(t)}{dt} = -\sum_{j=k}^{n} C_n^j.\frac{d}{dt}(R(t)^j.(1-R(t))^{n-j})$$

namely:

$$f_S(t) = -\sum_{j=k}^{n} C_n^j.\left[-j.f(t).R(t)^{j-1}.\left(1-R(t)\right)^{n-j} + R(t)^j.(n-j).\left(1-R(t)\right)^{n-j-1}.\left(-f(t)\right)\right]$$

That is:

$$f_S(t) = f(t).\sum_{j=k}^{n} C_n^j.R(t)^{j-1}.(1-R(t))^{n-j-1}[j.(1-R(t)) - R(t).(n-j)]$$

namely:

$$f_S(t) = f(t).\sum_{j=k}^{n} C_n^j.R(t)^{j-1}.(1-R(t))^{n-j-1}.(j - n.R(t)) \qquad [3.46]$$

Equations [3.31] and [3.46] indicate that the Rocof does not have explicit solutions. The asymptotic value of the Rocof is obtained from equations [3.24] and [3.34], that is:

$$Rocof_\infty = \frac{1}{\sum_{j=k}^{n}\sum_{i=0}^{n-j} C_n^j.C_{n-j}^i.(-1)^i.\int_0^{+\infty} exp(-(i+j).H(t)).dt} \qquad [3.47]$$

Exponential distribution

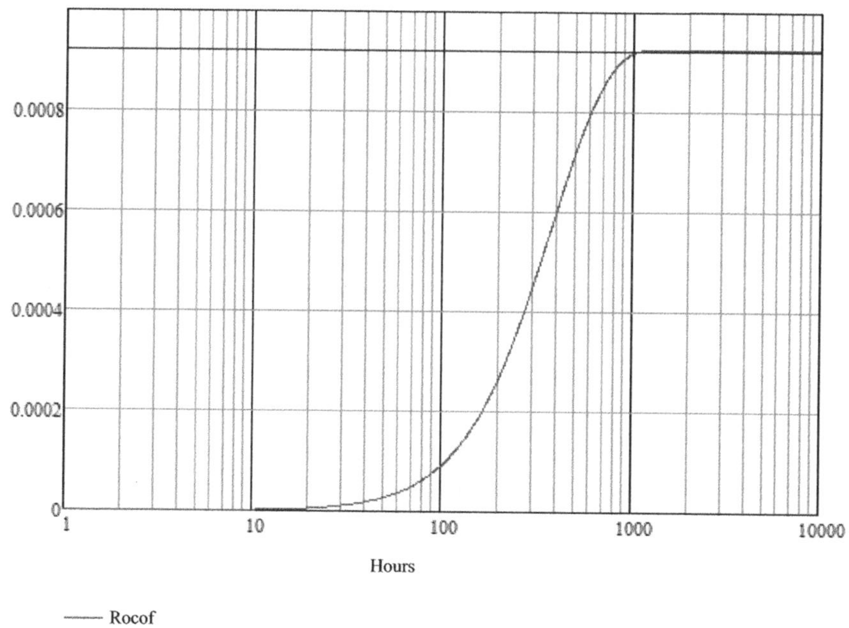

Figure 3.29. *Rocof of exponential distributions in k/n redundancy with maintenance at system level*

The Rocof has no explicit solutions. The Rocof asymptotic value is obtained from Appendix 1, namely:

$$Rocof_{\infty} = \frac{1}{MTTF_e \cdot \sum_{i=k}^{n} \frac{1}{i}} \qquad [3.48]$$

EXAMPLE 3.9.– Exponential distribution λ = 0.001 n = 4 and k = 2.

Simulation

We obtain the following figure in accordance with the one given by the numerical solution.

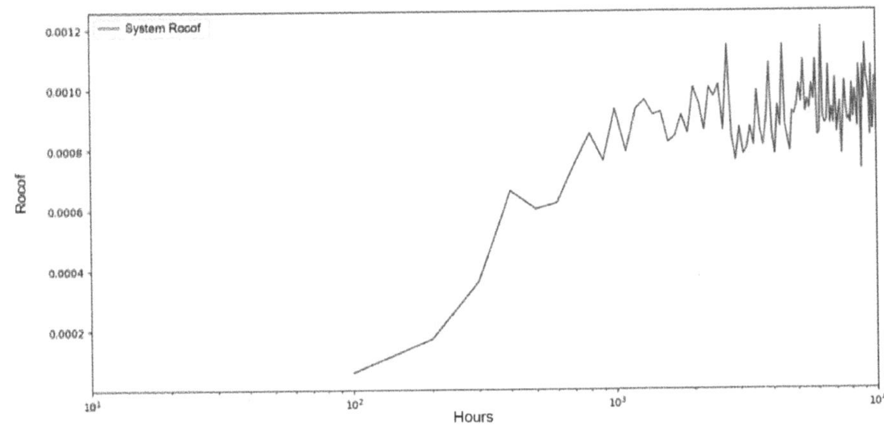

Figure 3.30. *Simulation of Rocof of exponential distributions in k/n redundancy with maintenance at system level*

Weibull distribution

The Rocof has no explicit solutions.

The Rocof asymptotic value is obtained from equations [3.34] and [3.27], that is:

$$Rocof_{\infty} = \frac{1}{\mathrm{MTTF}_W \cdot \sum_{j=k}^{n} \sum_{i=0}^{n-j} \frac{C_n^j \cdot C_{n-j}^i \cdot (-1)^i}{(i+j)^{1/\beta}}} \qquad [3.49]$$

EXAMPLE 3.10.– Weibull distribution $\eta = 100$; $\beta = 7$; n = 4 and k = 2.

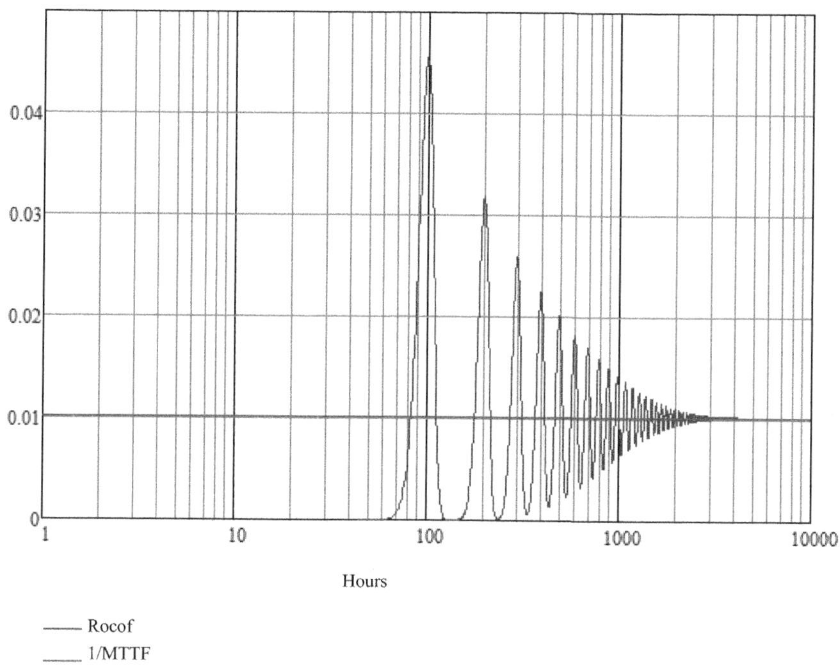

Figure 3.31. *Rocof of Weibull distributions in k/n redundancy with maintenance at system level*

Simulation

We obtain Figure 3.32, which is in agreement with the one given by the numerical solution.

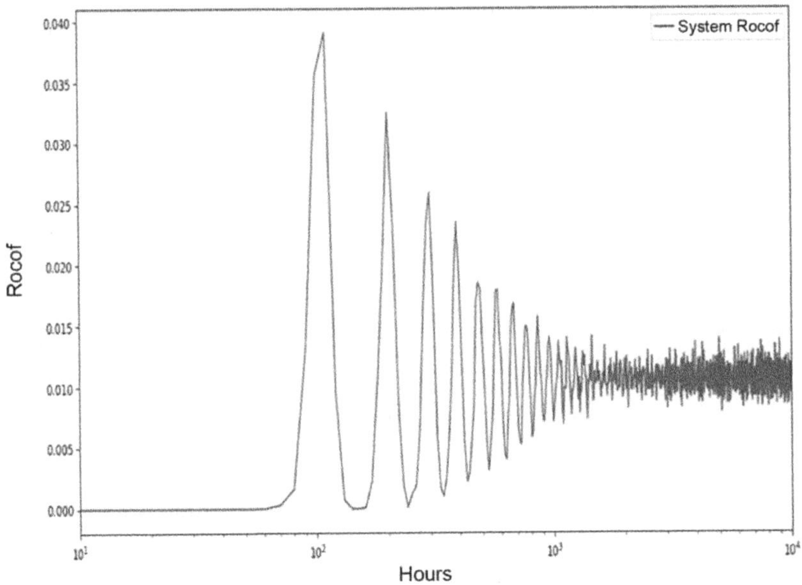

Figure 3.32. *Simulation of Rocof of Weibull distributions in k/n redundancy with maintenance at system level*

3.3.4.2.1. Summary

– *Element-level renewal*:

- Element-level renewal is identical to system renewal since all of the elements must be defective for the system to be defective.

– *System-level renewal*:

- The Rocof is obtained from the system probability density. It has no explicit solutions even if all of the elements follow an exponential distribution.

- For an exponential distribution, the asymptotic value is equal to $\frac{1}{\text{MTTF}_e \cdot \sum_{j=k}^{n} \frac{1}{j}}$ where MTTF is the MTTF of an exponential distribution.

- For a Weibull distribution, the asymptotic value is equal to $\frac{1}{\text{MTTF}_W \cdot \sum_{j=k}^{n} \frac{C_n^j \cdot (-1)^{j+1}}{j^{1/\beta}}}$ where MTTFw is the MTTF of a Weibull distribution.

3.4. Series/parallel system

These systems are made up of "s" subsets (S/E) in series. Each subset consists of a number of elements mounted in parallel that are unique to them but identical for each subset. Therefore, for subset "i", the number of elements is denoted as "p_i". Such systems can be illustrated as follows if element i,j represents the *i*th element of subset j.

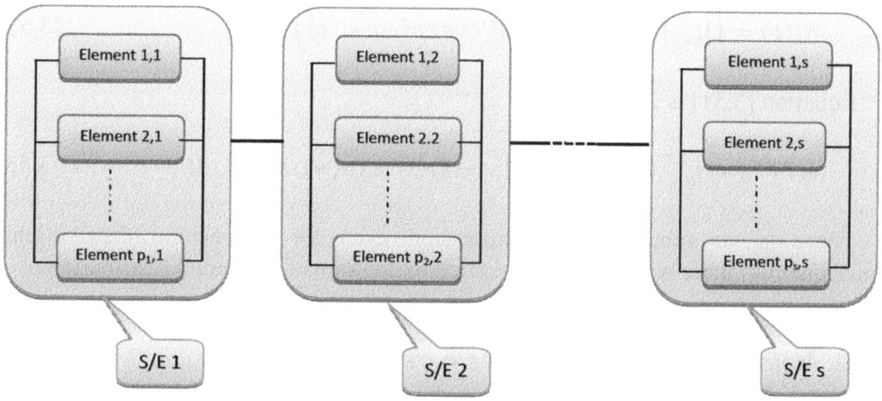

Figure 3.33. *Illustration of a series/parallel system*

3.4.1. *Maintenance-free serial/parallel system*

The system will then be reliable if all of the "s" serial subsets are also reliable. The reliability function of the serial system is given by:

$$R_S(t) = \prod_{i=1}^{s} R_{S/E_i}(t)$$

The subset "i" is defective if the p_i elements that it comprises are also defective since they are in parallel. Its reliability function is given by equation [3.16], that is:

$$R_{S/E_i}(t) = \sum_{j=1}^{p_i} C_{p_i}^{j}.(-1)^{j+1}.exp(-j.H_i(t))$$

from which:

$$R_S(t) = \prod_{i=1}^{s}\left(\sum_{j=1}^{p_i} C_{p_i}^{j}.(-1)^{j+1}.exp(-j.H_i(t))\right) \quad [3.50]$$

The system MTTF is therefore given by:

$$MTTF_S = \int_0^{+\infty} \prod_{i=1}^{s} \left(\sum_{j=1}^{p_i} C_{p_i}^j . (-1)^{j+1} . exp(-j.H_i(t)) \right) . dt \qquad [3.51]$$

Exponential distributions

We have here $H(t) = \lambda . t$; thus, equation [3.50] can be written as:

$$R_S(t) = \prod_{i=1}^{s} \left(\sum_{j=1}^{p_i} C_{p_i}^j . (-1)^{j+1} . exp(-j.\lambda_i.t) \right) \qquad [3.52]$$

Equation [3.51] is as follows:

$$MTTF_S = \int_0^{+\infty} \prod_{i=1}^{s} \left(\sum_{j=1}^{p_i} C_{p_i}^j . (-1)^{j+1} . exp(-j.\lambda_i.t) \right) . dt \qquad [3.53]$$

There is no simple MTTF expression even for exponential distributions. Consequently, there will also be no explicit expressions for Weibull distributions.

3.4.2. *Maintenance-enabled series/parallel systems*

3.4.2.1. *Element-level renewal*

To further illustrate the reasoning that was used, consider a system with two subsets in series, each of which has two identical elements assembled in parallel.

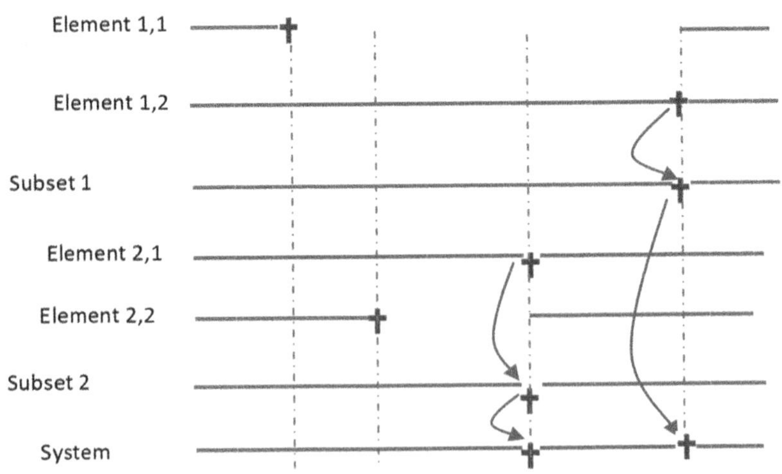

Figure 3.34. *Illustration of failures in an element-renewal "series/parallel" system*

At the "subset" level, we find a simple renewal process, as we observed in section 3.3.3, since the elements are in parallel.

Therefore, at the "system" level, given that the subsets are in series, we will see a superposition of renewal processes, as described in section 3.3.4. Only the expressions of the probability densities of the subsets will be different.

We can therefore write that, for the subset "i":

$$F_{S/E_i}(t) = F_i^{p_i}(t) \text{ from which } f_{S/E_i}(t) = p_i.F_i^{p_i-1}(t).f_i(t)$$

The system Rocof is therefore written as:

$$Rocof_{SP}(t) = \sum_{i=1}^{S} Rocof_{S/E_i}(t) = \sum_{i=1}^{S}\sum_{j=1}^{+\infty}\left(f_{S/E_i}(t)\right)^{<j>}$$

or still:

$$Rocof_{SP}(t) = \sum_{i=1}^{S}\sum_{j=1}^{+\infty}\left(p_i.F^{p_i-1}(t).f(t)\right)^{<j>} \quad [3.54]$$

The asymptotic value of the Rocof is obtained from equations [3.26] and [3.34], that is:

$$Rocof_{\infty} = \frac{1}{\int_0^{+\infty}\prod_{i=1}^{S}\left(\sum_{j=1}^{p_i} C_{p_i}^{j}.(-1)^{j+1}.exp(-j.H_i(t))\right).dt} \quad [3.55]$$

Exponential distribution

Here, we have:

$$f_{S/E_i}(t) = p_i.\left(1 - exp(-\lambda_i.t)\right)^{p_i-1}.\lambda_i.exp(-\lambda_i.t)$$

Using Newton's binomial formula, we obtain:

$$f_{S/E_i}(t) = p_i.\lambda_i.exp(-\lambda_i.t).\sum_{k=0}^{p_i-1} C_{p_i-1}^{k}.(-1)^k.exp(-\lambda_i.t)^k$$

or still:

$$f_{S/E_i}(t) = p_i.\lambda_i.\sum_{k=0}^{p_i-1} C_{p_i-1}^{k}.(-1)^k.exp(-(k+1).\lambda_i.t)$$

The system Rocof is therefore equal to:

$$Rocof_{SP}(t) = \sum_{i=1}^{s}\sum_{j=1}^{+\infty}\left(p_i.\lambda_i.\sum_{k=0}^{p_i-1} C_{n_i-1}^{k}.(-1)^k.exp(-(k+1).\lambda_i.t)\right)^{<j>} \quad [3.56]$$

Even for exponential distributions, the Rocof has no analytical solutions.

The asymptotic value of the Rocof for the series/parallel system is given by, since we have "s" subsets in series (equation [3.41]):

$$\text{Rocof}_\infty = \sum_{i=1}^{s} \text{Rocof}_{\infty_i}(t)$$

However, the subsets are parallel systems such that each of them is a simple renewal process whose asymptotic value is given by equation [3.33]:

$$\text{Rocof}_\infty = \sum_{i=1}^{s}\left(\frac{\lambda_i}{\sum_{j=1}^{p_i}\frac{1}{j}}\right) \quad [3.57]$$

Weibull distributions

The Rocof has no explicit solutions since there are no solutions for the exponential distribution.

The asymptotic Rocof is obtained from equation [3.34]. Namely:

$$\text{Rocof}_\infty = \sum_{i=1}^{s}\left(\frac{p_i.\Gamma\left(1+\frac{1}{\beta_i}\right)}{\sum_{j=1}^{p}\frac{C_{p_i}^{j}.(-1)^{j+1}}{j^{1/\beta_i}}}\right) \quad [3.58]$$

3.4.3. *System-level renewal*

Here, as soon as a subset fails, the whole system is replaced by a new one, as illustrated in Figure 3.35.

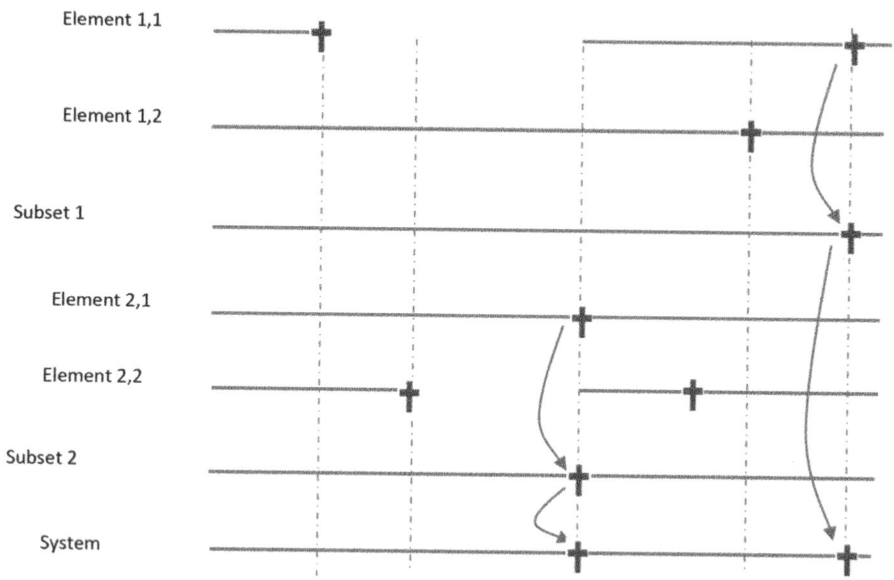

Figure 3.35. *Serial/parallel renewal system without element test*

The system reliability function can therefore be written, since we have "s" elements in series, as:

$$R_{SP}(t) = \prod_{i=1}^{s} R_{S/E_i}(t)$$

According to Appendix 3, we can write the following:

$$f_{SP}(t) = \sum_{i=1}^{s} \left(f_{S/E_i}(t). \prod_{\substack{k=1 \\ k \neq i}}^{s} R_{S/E_k}(t) \right)$$

The reliability function of the "*j*th" subset with p_k elements is obtained from equation [3.16], that is:

$$f_{SP}(t) = \sum_{i=1}^{s} \left(p_i . F_i^{p_i}(t). f_i(t). \prod_{\substack{k=1 \\ k \neq i}}^{s} \left(\sum_{j=1}^{p_k} C_{p_k}^{j} . (-1)^{j+1}. H_k(t) \right) \right) \qquad [3.59]$$

The Rocof therefore has no explicit solutions. The MTTF is given by:

$$MTTF_{SP} = \int_0^{+\infty} \prod_{i=1}^{s} \left(\sum_{j=1}^{p_i} C_{p_i}^{j} . (-1)^{j+1} . R_i(t)^j \right) . dt$$

or still:

$$MTTF_{SP} = \int_0^{+\infty} \prod_{i=1}^{p} \left(\sum_{j=1}^{n_i} C_{n_i}^{j} . (-1)^{j+1} . exp(-j . H_i(t)) \right) . dt \qquad [3.60]$$

3.4.4. *Summary*

– *System-level renewal*:

- The Rocof is obtained from the system probability density. It has no explicit solutions even if all of the elements follow an exponential distribution.

- The asymptotic value of the Rocof also has no explicit solutions.

– *Subset-level renewal*:

- Given that we have a serial system, the system Rocof is equal to the sum of the Rocof of each subset. Since each subset is a parallel system, subset renewal is identical to element renewal.

– *Element-level renewal*:

- We do not know of any theoretical solutions to address this case.

– *For the three maintenance strategies, it is therefore necessary to use numerical computations or Monte Carlo simulations.*

3.5. Parallel/serial system

These systems include “p” identical subsets (S/E) in parallel. Therefore, all of the subsets comprise “s” elements in series.

These systems can be depicted in Figure 3.36.

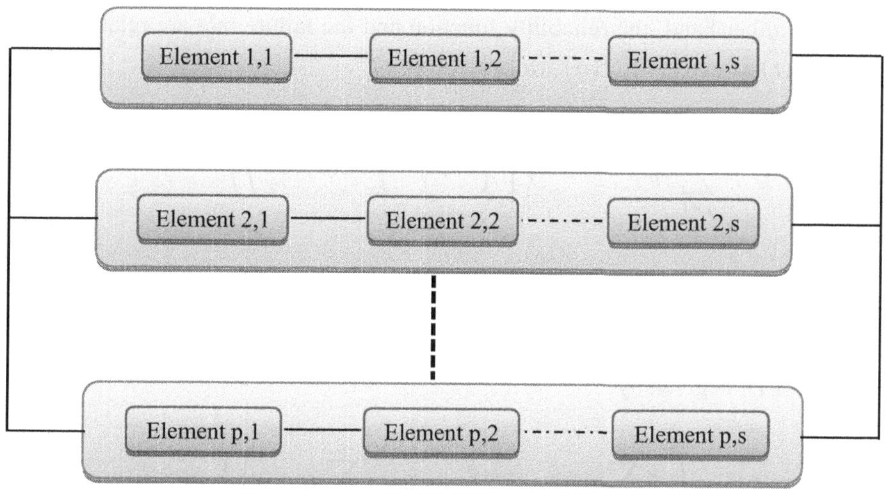

Figure 3.36. *Illustration of a parallel/serial system*

3.5.1. *Maintenance-free parallel/serial system*

The reliability function of the subset "k", consisting of "s" elements in series, is given by:

$$R_k(t) = \prod_{i=1}^{s} R_i(t) \text{ from which } F_k(t) = 1 - \prod_{i=1}^{s} R_i(t)$$

REMARK.–

Strictly speaking, we should write $\mathrm{R_k(t)} = \prod_{i=1}^{S_k} \mathrm{R_i(t)}$ from which $\mathrm{F_k(t)} = 1 - \prod_{i=1}^{S_k} \mathrm{R_{k,i}(t)}$, but since all subsets in parallel are identical, the previous notation is also valid.

The reliability function of the "parallel/series" system can therefore be written, as the "p" subsets in parallel are identical:

$$R_{PS}(t) = 1 - F_{PS}(t) = 1 - \left(1 - \prod_{i=1}^{s} R_i(t)\right)^p$$

This starts from the formula of the Newton binomial:

$$R_{PS}(t) = \sum_{j=1}^{p} C_p^j.(-1)^{j+1}.\left(\prod_{i=1}^{s} R_i(t)\right)^j$$

On the other hand, the reliability function and the failure rate are related by the relation $R(t) = \exp\left(-\int_0^t \lambda(u).du\right)$:

$$R_{PS}(t) = \sum_{j=1}^{p} C_p^j.(-1)^{j+1}.\left(\prod_{i=1}^{s} \exp\left(-\int_0^t \lambda_i(u).du\right)\right)^j$$

that is:

$$R_{PS}(t) = \sum_{j=1}^{p} C_p^j.(-1)^{j+1}.exp(-j.\sum_{i=1}^{s} H_i(u)) \qquad [3.61]$$

The MTTF is given by:

$$MTTF_{PS} = \int_0^{+\infty} \left(\sum_{j=1}^{p} C_p^j.(-1)^{j+1}.exp\left(-j.\sum_{i=1}^{s} H_i(u)\right)\right).dt$$

namely:

$$MTTF_{PS} = \sum_{j=1}^{p} C_p^j.(-1)^{j+1}.\int_0^{+\infty} exp(-j.\sum_{i=1}^{s} H_i(t)).dt \qquad [3.62]$$

Exponential distribution

The failure rate is constant such that:

$$R_{PS}(t) = \sum_{j=1}^{p} C_p^j.(-1)^{j+1}.\exp(-j.\sum_{i=1}^{s} \lambda_i.t) \qquad [3.63]$$

On the other hand, according to the previous equation, the following can be written:

$$MTTF_{SP} = \sum_{j=1}^{p} C_p^j.(-1)^{j+1}.\int_0^{+\infty} exp\left(-j.\sum_{i=1}^{s} \lambda_i.t\right).dt$$

or, finally, by setting $u = j.\sum_{i=1}^{s} \lambda_i.t$:

$$MTTF_{SP} = \frac{1}{\sum_{i=1}^{s} \lambda_i}.\sum_{j=1}^{p} \frac{C_p^j.(-1)^{j+1}}{j}$$

or, still according to Appendix 1:

$$MTTF_{SP} = \frac{1}{\sum_{i=1}^{s} \lambda_i}.\sum_{j=1}^{p} \frac{1}{j} \qquad [3.64]$$

REMARK.–

From equation [3.11], the previous equation can also be written as $\mathrm{MTTF_{SP}} = \mathrm{MTTF_S}.\sum_{j=1}^{p}\frac{1}{j}$, which is the parallelization of "p" identical elements.

Weibull distributions

The preceding remark allows us to write:

$$MTTF_{SP} = MTTF_{S/E}.\sum_{j=1}^{p}\frac{1}{j}$$

Equation [3.13] leads to:

$$MTTF_{SP} = \int_0^{+\infty} exp\left(-\sum_{i=1}^{s} exp\left(\left(\frac{t}{\eta_i}\right)^{\beta_i}\right)\right).\sum_{j=1}^{p}\frac{1}{j} \quad [3.65]$$

3.5.2. *Maintenance-enabled parallel/serial system*

3.5.2.1. *Element-level renewal*

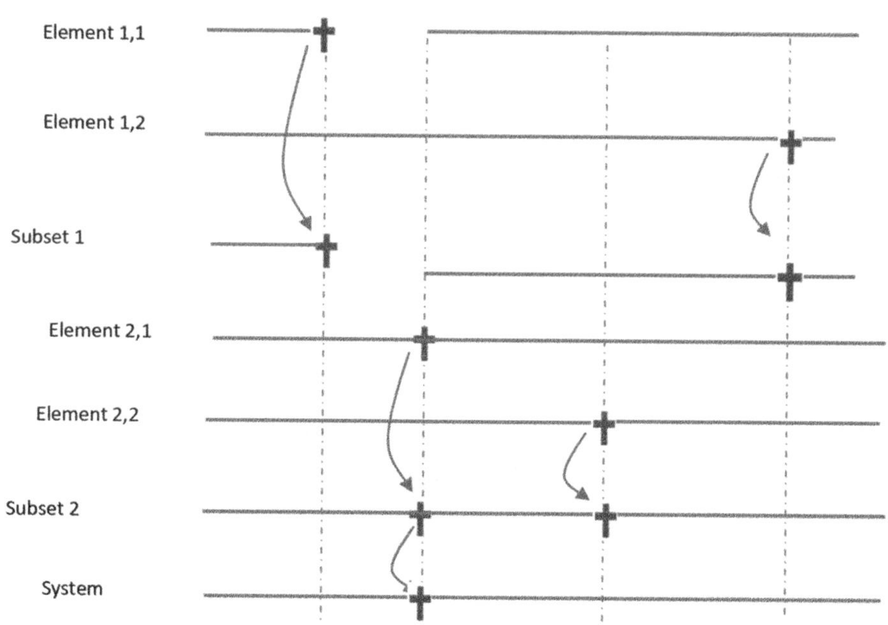

Figure 3.37. *Illustration of failures in a "parallel/series" element-level renewal system*

When an element fails, since it is not tested, it causes the failure of the subset to which it belongs.

However, because the subsets are in parallel, this does not cause system failure.

Therefore, the system will be faulty when all of the subsets are faulty, as shown below for two subsets and two items per subset.

The system probability density is given by:

$$f_{PS}(t) = -\frac{\mathrm{d}R_{PS}(t)}{\mathrm{dt}} = -\frac{\mathrm{d}}{\mathrm{dt}}\left(\sum_{j=1}^{p} C_p^j.(-1)^{j+1}.exp\left(-j.\sum_{i=1}^{s} H_i(t)\right)\right)$$

that is:

$$f_{PS}(t) = -\sum_{j=1}^{p} C_p^j.(-1)^{j+1}.\frac{\mathrm{d}}{\mathrm{dt}}\left(exp\left(-j.\sum_{i=1}^{s} H_i(t)\right)\right)$$

namely:

$$f_{PS}(t) = \sum_{j=1}^{p} C_p^j.(-1)^{j+1}.exp(-j.\sum_{i=1}^{s} H_i(t)).(j.\sum_{i=1}^{s} \lambda_i(t)) \qquad [3.66]$$

Exponential distributions

$\lambda_i(t) = \lambda_i$ and $H_i(t) = \lambda_i.t$, from which:

$$f_{PS}(t) = \sum_{j=1}^{p} C_p^j.(-1)^{j+1}.exp\left(-j.\sum_{i=1}^{s} \lambda_i.t\right).\left(\sum_{i=1}^{s} \lambda_i\right)$$

EXAMPLE 3.11.– With p = 5; s = 2; $\lambda 1 = 1/200$; $\lambda 2 = 1/300$.

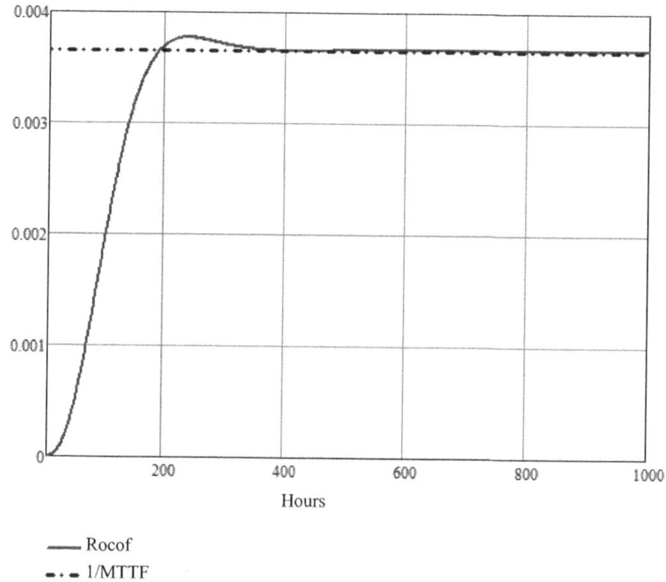

Figure 3.38. *Rocof of exponential distributions for parallel/ series system with maintenance at element level*

Simulation

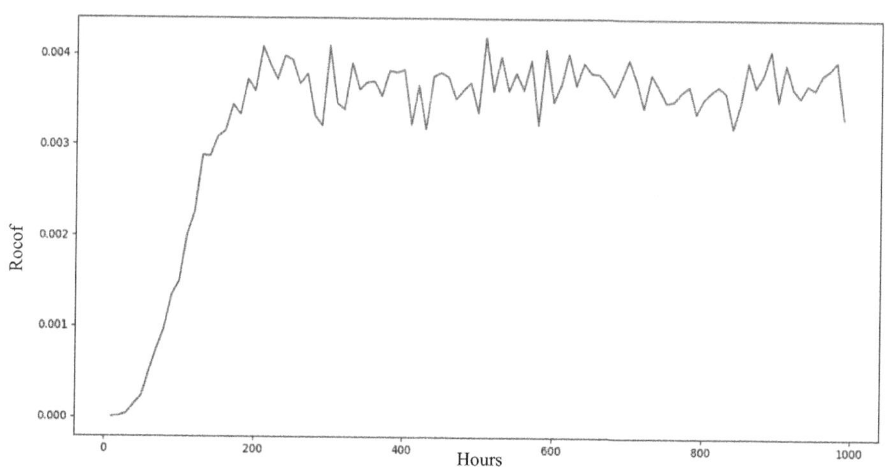

Figure 3.39. *Simulation of Rocof of exponential distributions for parallel/series system with maintenance at element level*

Weibull distributions

$\lambda_i(t) = \left(\frac{\beta_i}{\eta_i}\right).\left(\frac{t}{\eta_i}\right)^{\beta_i-1}$ and $H_i(t) = \left(\frac{t}{\eta_i}\right)^{\beta_i}$ from which:

$$f_{PS}(t) = \sum_{j=1}^{p} C_p^j.(-1)^{j+1}.exp\left(-j.\sum_{i=1}^{s}\left(\frac{t}{\eta_i}\right)^{\beta_i}\right).\left(j.\sum_{i=1}^{s}\left(\frac{\beta_i}{\eta_i}\right).\left(\frac{t}{\eta_i}\right)^{\beta_i-1}\right)$$

EXAMPLE 3.12.– p = 5; s = 2; η1 = 200; β1 = 1.5; η2 = 300; β2 = 4.

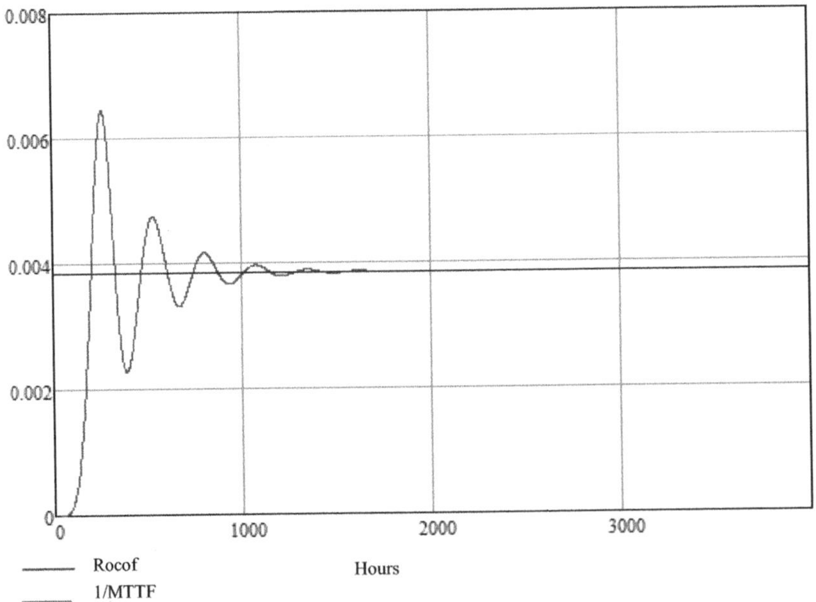

Figure 3.40. *Parallel/series system Rocof Weibull distributions element renewal. For a color version of this figure, see www.iste.co.uk/bayle/montecarlo.zip*

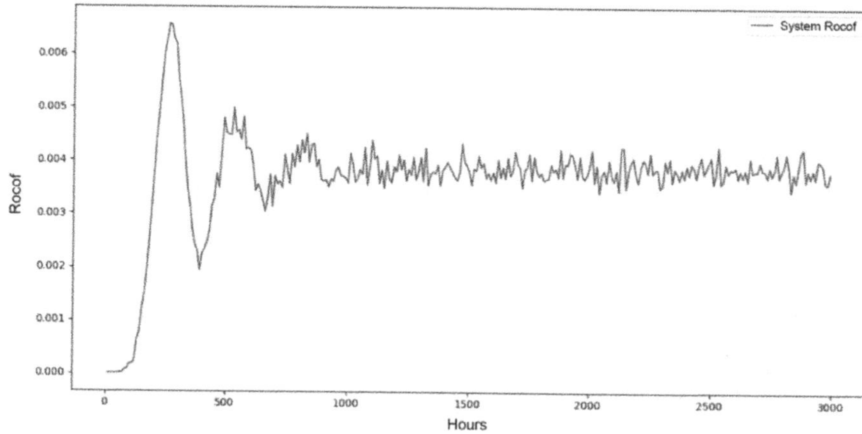

Figure 3.41. *Rocof of Weibull distributions for parallel/series system with maintenance at element level*

3.5.2.2. *System-level renewal*

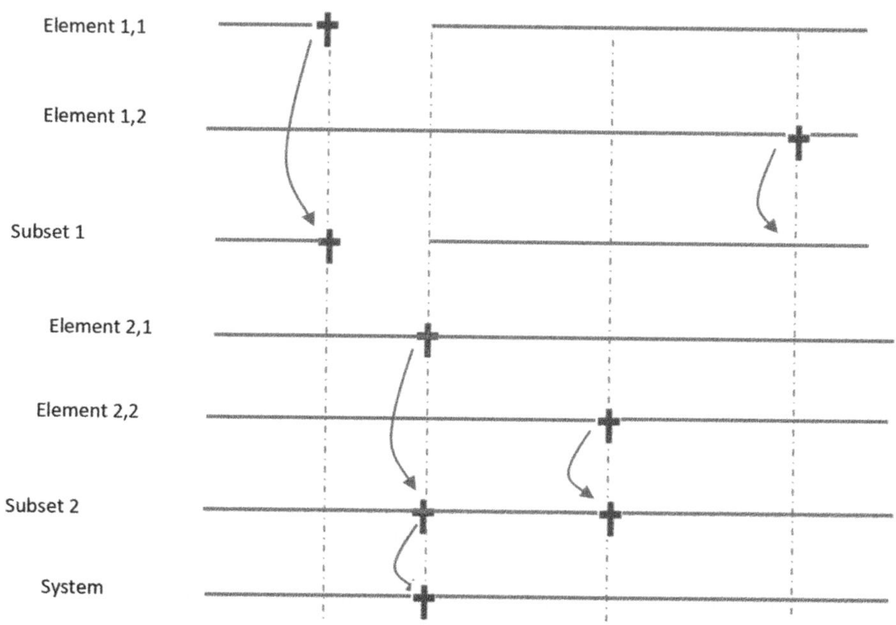

Figure 3.42. *Illustration of a parallel/serial system: system renewal*

For the system to fail, all of the subsets must be faulty. This strategy replaces all the elements of all the subsets with new ones, even if most of them have not failed. This can be illustrated by Figure 3.42.

3.5.2.3. *Summary*

– *System-level renewal*:

- The Rocof is obtained from the system probability density. It has no explicit solutions even if all of the elements follow an exponential distribution.

- The asymptotic value of the Rocof also has no explicit solutions.

– *Subset-level renewal*:

- Since we are considering a parallel system, renewal at the subset level is identical to system renewal.

– *Element-level renewal*:

- Bearing in mind that the subsets are made up of series elements, the Rocof of each subset is equal to the sum of the Rocof of each element that it contains. The system fails as soon as at least one of the elements of each subset fails. We do not know of any analytical solutions to address this case.

– *For the three maintenance strategies, numerical computations or Monte Carlo simulations have thus to be employed.*

3.6. Use cases

3.6.1. *Thermostatic energy conversion*

Consider a system performing a high-power energy conversion. The core power of such systems is achieved by discrete power components such as MOSFET or IGBT transistors operating in switching mode and capable of yielding fairly high power efficiency. Nevertheless, they are subject to losses either due to the Joule effect (when the component is On) or the switching mode (when switching On/Off). These losses induce an increase in temperature, not only at the transistor level but also inside the system. To achieve acceptable junction temperatures in these power components, a fan is inserted into the system.

In addition, to increase system reliability, it was decided that for project 2, identical fans had to be implemented in parallel. We then obtain the following diagram in Figure 3.43.

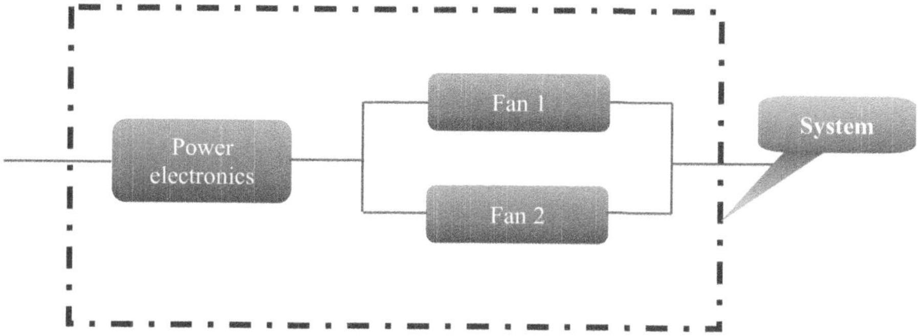

Figure 3.43. *Equivalent diagram of the energy conversion system*

We see that this is a *series/parallel system*, and the first subset (the electronics) has no elements in parallel. The electronics can be modeled by an exponential distribution, whereas the fan, which undergoes an aging mechanism, is modeled by a Weibull distribution.

REMARK.–

In this calculation, we did not consider the effect of the system temperature. In practice, when a fan fails, the internal temperature of the system increases, which obviously affects the reliability level of the electronics and the other fan. On the other hand, we also assume that there is no monitoring of the fans. In practice, some fans have an image output of the fan rotation speed so that its failure (the fan no longer rotates) can be immediately detected.

3.6.1.1. *Element-level renewal*

Given that this is a serial system including the electronics and the two fans all together, if the system is functional, it means that the electronics and the two fans are not faulty. It is very likely that, considering the entire array of operational systems, there is a system in which the electronics fail first. Since the system is in series, it fails. The repair will then consist of replacing the electronics.

A fan will fail, but since they are in parallel, it will not have any effect on the system level. Next, the second fan will fail causing the system to fail. The repair will then consist of replacing the two faulty fans, and the scenario will be repeated in this way.

The electronics and the two fans are in series; therefore, we are dealing with a superposition of two renewal processes. We can therefore write the following:

$$N_s(t) = N_{Elec}(t) + N_{2fan}(t)$$

where:

– N_{elec} is the number of electronic failures at time "t";

– N_{2fan} is the number of failures of the two fans.

We thus have:

$$Rocof_s(t) = \frac{\delta N_s(t)}{\delta t} = Rocof_{Elec}(t) + Rocof_{2fan}(t)$$

Since electronics are modeled by an exponential distribution, we have:

$$Rocof_{Elec}(t) = \lambda.$$

However, the probability density of the two fans in parallel can be written as:

$$f_{2fan}(t) = \frac{\mathrm{dF2fan}(t)}{\mathrm{dt}} = \frac{\mathrm{d}}{\mathrm{dt}}\left(1 - exp\left(-\left(\frac{t}{\eta}\right)^{\beta}\right)\right)^2$$

that is:

$$f_{2fan}(t) = 2.\left(\frac{\beta}{\eta}\right).\left(\frac{t}{\eta}\right)^{\beta-1}.exp\left(-\left(\frac{t}{\eta}\right)^{\beta}\right).\left(1 - exp\left(-\left(\frac{t}{\eta}\right)^{\beta}\right)\right)$$

Consequently, the system Rocof is given by:

$$Rocof_S = \lambda + \sum_{i=1}^{+\infty}\left(2.\left(\frac{\beta}{\eta}\right).\left(\frac{t}{\eta}\right)^{\beta-1}.exp\left(-\left(\frac{t}{\eta}\right)^{\beta}\right).\left(1 - exp\left(-\left(\frac{t}{\eta}\right)^{\beta}\right)\right)\right)^{<i>}$$

The asymptotic value of the system Rocof is obtained from equation [3.36], namely:

$$\mathrm{Rocof}_{\infty} = lim_{t\to\infty}\mathrm{Rocof}_{elec} + lim_{t\to\infty}\mathrm{Rocof}_{2fan}$$

that is:

$$\mathrm{Rocof}_{\infty} = \lambda + \frac{1}{MTTF_{2fan}}$$

However, $R_{2fan}(t) = 2.exp\left(-\left(\frac{t}{\eta}\right)^{\beta}\right) - exp\left(-2.\left(\frac{t}{\eta}\right)^{\beta}\right)$ from which:

$$MTTF_{2fan} = \int_0^{+\infty} R_{2fan}(t).dt = \int_0^{+\infty} 2.exp\left(-\left(\frac{t}{\eta}\right)^{\beta}\right) - exp\left(-2.\left(\frac{t}{\eta}\right)^{\beta}\right).dt$$

that is:

$$MTTF_{2fan} = \left(2 - 2^{-1/\beta}\right).\eta.\Gamma\left(1+\frac{1}{\beta}\right) = MTTF_{fan}.\left(2 - 2^{-1/\beta}\right)$$

and thus:

$$\text{Rocof}_{\infty} = \lambda + \frac{1}{\left(2 - 2^{-1/\beta}\right).\eta.\Gamma\left(1+\frac{1}{\beta}\right)}$$

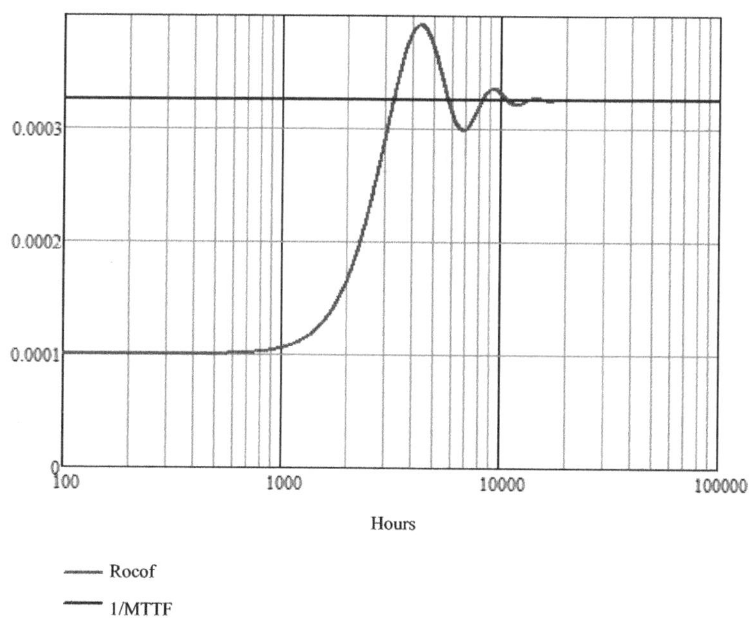

Figure 3.44. *Rocof of series/parallel system with maintenance at element level*

We then obtain the following figure if we have an exponential distribution of parameters $\lambda = 10^{-4}$ for the electronic components and a Weibull distribution of parameters $\eta = 4{,}000$ and $\beta = 2.4$ for the fans.

The asymptotic value of the Rocof is, according to the previous figure:

$$\text{Rocof}_{\infty} = 10^{-4} + \frac{1}{(2-2.4).40{,}000.\Gamma\left(1+\frac{1}{2.4}\right)} \approx 3.255 \times 10^{-4}$$

3.6.1.2. *Renewal at the subset level (two fans)*

Since the subset of fan comprises identical elements (the fans) in parallel, this strategy is equivalent to the previous one.

3.6.1.3. *System-level renewal*

We have a system renewal if the electronics are faulty or if the two fans are. The probability of system failure is given by:

$$F_S(t) = 1 - R_{elec}(t).R_{2fans}(t)$$
$$= exp(-\lambda.t).\left(2.exp\left(-\left(\frac{t}{\eta}\right)^{\beta}\right) - exp\left(-2.\left(\frac{t}{\eta}\right)^{\beta}\right)\right)$$

from which:

$$f_s(t) = \exp\left(-\lambda.t - \left(\frac{t}{\eta}\right)^{\beta}\right).\left(\left(\frac{2.\beta}{\eta}\right).\left(\frac{t}{\eta}\right)^{\beta-1}.\left(1 - exp\left(-\left(\frac{t}{\eta}\right)^{\beta}\right)\right)\right.$$
$$\left.- \lambda.\left(exp\left(-\left(\frac{t}{\eta}\right)^{\beta}\right) - 2\right)\right)$$

which allows us to compute the corresponding Rocof.

On the other hand, the system MTTF is given by:

$$MTTF_S = \int_0^{+\infty} R_S(t).dt$$
$$= \int_0^{+\infty} R_{elec}(t).R_{2fans}(t).dt$$
$$= \int_0^{+\infty} exp(-\lambda.t).\left(2.exp\left(-\left(\frac{t}{\eta}\right)^{\beta}\right)\right.$$
$$\left.- exp\left(-2.\left(\frac{t}{\eta}\right)^{\beta}\right)\right).dt$$

We therefore obtain Figure 3.45.

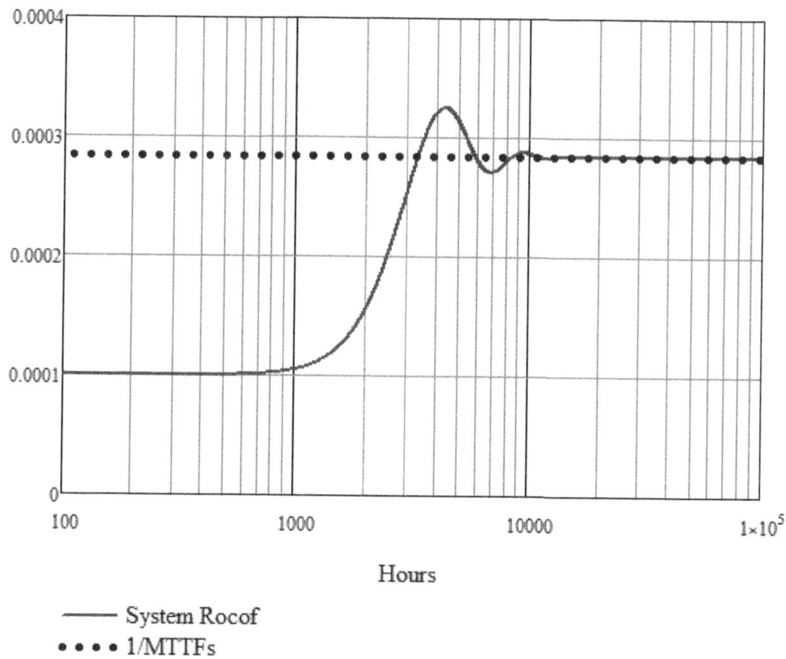

Figure 3.45. *Rocof of series/parallel system with maintenance at system level*

The asymptotic value of the Rocof is approximately 2.841×10^{-4}. It can be seen that it is lower for the system-level renewal strategy (2.841×10^{-4}) than for the subset-level renewal (3.255×10^{-4}), which makes sense.

3.6.2. *Avionics computer*

Consider a computer embedded in a civil aircraft consisting of serial electronics with a human–machine interface.

A packaging, with a failure rate that is considered zero, encloses everything. Pilot/copilot redundancy causes the previous subset to be assembled in parallel such that we obtain Figure 3.46.

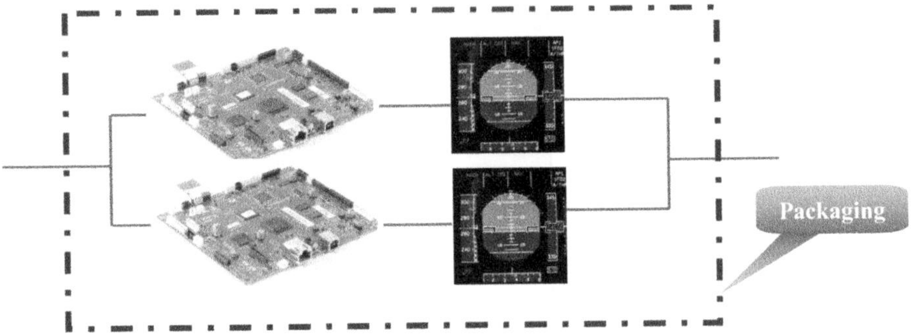

Figure 3.46. *Avionics system diagram*

This is therefore a parallel/series system with p = 2 and s = 2. The electronics can be modeled by an exponential distribution of parameter λ, whereas the human/machine interface can be modeled by a Weibull distribution of parameters η and β.

3.6.2.1. *Element-level renewal*

Element-level renewal leads to the following illustration:

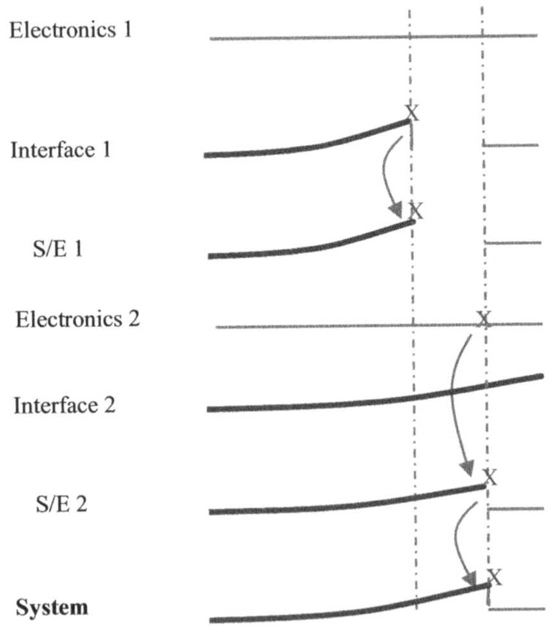

Figure 3.47. *Illustration of failures in an avionics system with element renewal*

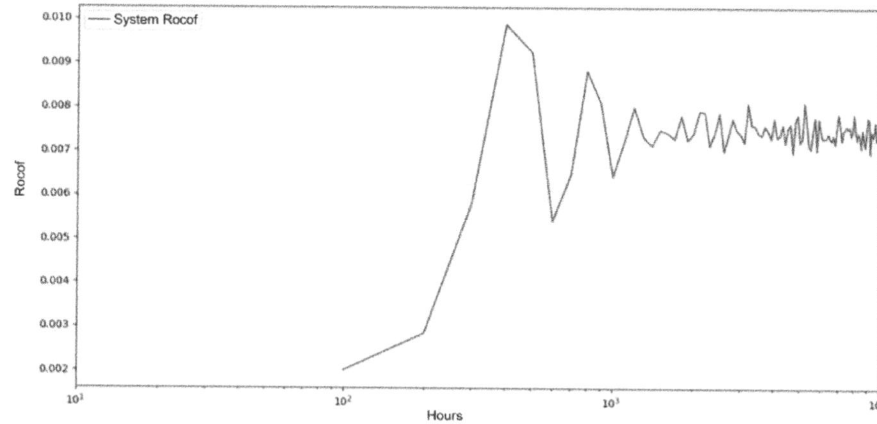

Figure 3.48. *Simulation of Rocof of avionic system with maintenance at element level*

3.6.2.2. *System-level renewal*

Here, it is no longer the element or subset that is renewed but the entire system when it fails.

From equation [3.55], we obtain:

$$f(t) = \sum_{j=1}^{2} C_2^j.(-1)^{j+1}.exp\left(-j.\sum_{i=1}^{2} H_i(t)\right).\left(j.\sum_{i=1}^{2} \lambda_i(t)\right)$$

Namely:

$$f(t) = 2.exp\left(-\lambda.t - \left(\frac{t}{\eta}\right)^{\beta}\right).\left(\lambda + \left(\frac{t}{\eta}\right)^{\beta-1}\right) - exp\left(-2.\left(-\lambda.t - \left(\frac{t}{\eta}\right)^{\beta}\right)\right).\left(2.\left(\lambda + \left(\frac{t}{\eta}\right)^{\beta-1}\right)\right)$$

and the MTTF is obtained from equation [3.50], that is:

$$MTTF = \sum_{j=1}^{2} C_p^j.(-1)^{j+1}.\int_0^{+\infty} exp\left(-j.\sum_{i=1}^{2} H_i(t)\right).dt$$

namely:

$$MTTF = 2.\int_0^{+\infty} exp\left(\lambda.t + \left(\frac{t}{\eta}\right)^{\beta}\right).dt - \int_0^{+\infty} exp\left(-2.\left(\lambda.t + \left(\frac{t}{\eta}\right)^{\beta}\right)\right).dt$$

By taking λelec = 0.001, η = 400, and β = 4.7, we obtain the Rocof shown in Figure 3.49.

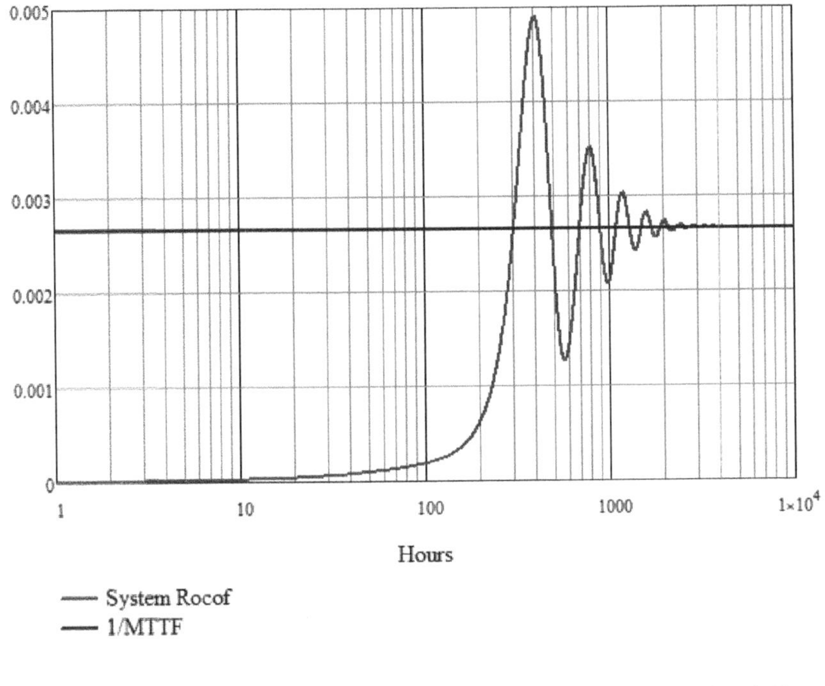

Figure 3.49. *Rocof of avionic system with maintenance at system level. For a color version of this figure, see www.iste.co.uk/bayle/montecarlo.zip*

Simulation

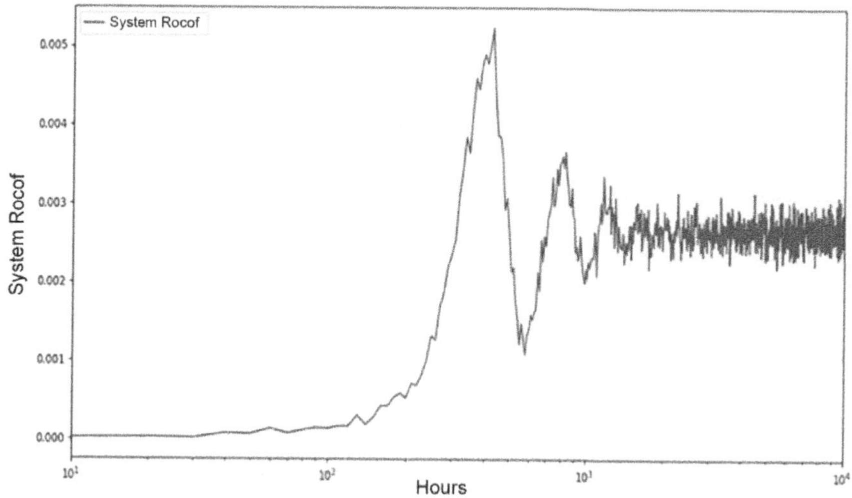

Figure 3.50. *Simulation of Rocof of avionic system with maintenance at system level*

REMARK.–

In practice, from a safety point of view, this case will occur if system failure causes a critical event. To limit the probability of this event to a value specified by the principal, all of the system subsets are tested before each operational mission. If, during this test, system failure is observed, the system will be changed. This point will be covered in detail in Part 4.

4

Impact of Temperature on Reliability

4.1. Arrhenius law

Toward the end of the 19th century, the Swedish chemist Arrhenius discovered, while studying chemical reactions on sugarcane, that the rate of chemical reaction increases when the temperature increases. He even proposed a model, the famous Arrhenius law (Arrhenius 1889), which defines the rate of chemical reaction as a function of temperature, that is:

$$Rate = \mathrm{A.exp}\left(\frac{\mathrm{Ea}}{\mathrm{Kb.}\,\theta}\right) \qquad [4.1]$$

where:

– Ea is the activation energy;

– Kb is the Boltzmann constant;

– A is the constant known as a "preexponential" constant;

– θ is the temperature in Kelvin.

The Arrhenius law is therefore an empirical law based on the results observed experimentally in most cases.

In the middle of the 20th century, Schockley reported the same trend in transistor reliability and proposed the use of this law to model the effect of temperature in terms of reliability. The purpose of this chapter is not to discuss the validity of this law. We refer the reader to Lall et al. (1997).

For a color version of all of the figures in this chapter, see www.iste.co.uk/bayle/montecarlo.zip.

4.2. Operational life profile

When assessing the predictive reliability of a system or when estimating the reliability of a component that has undergone accelerated testing, the notion of a life profile is a major element (FIDES 2022). The temperature levels in most life profiles can vary from one operational phase to another such that the Arrhenius law is not directly applicable. As an example, the different life profiles detailed in the FIDES (2022) guide can be mentioned here and are illustrated for a helicopter in Table 4.1.

Helicopter - On-board navigation computer - VIP use														
		Temperature and humidity			Thermal cycling				Mechanical	Chemical				Induced
Phase title	Calendar time (hours)	On/Off	Ambient temperature (°C)	Humidity level (%)	ΔT (°C)	Number of cycles (/an)	Cycle time (hours)	Maximal temperature during cycling (°C)	Random vibration (Grms)	Salt pollution	Environmental pollution	Application pollution	Protection level	Π application
Off -24h	6360	Off	15	70	10	265	24	20	0.01	Low	Moderate	Moderate	Non-hermetic	2.33
Off	2146	Off	15	70	10	100	20,6	20	0.01	Low	Moderate	Moderate	Non-hermetic	2.33
Sol-On 1	27	On	36	20	30	100	0.6	45	0.5	Low	Moderate	Moderate	Non-hermetic	2.71
Vol 1	100	On	35	20	10	100	1.1	45	6	Low	Moderate	Moderate	Non-hermetic	2.51
Sol-On2	27	On	36	20	30	100	0.6	45	0.5	Low	Moderate	Moderate	Non-hermetic	2.71
Vol 2	100	On	35	20	10	100	1.1	45	6	Low	Moderate	Moderate	Non-hermetic	2.51

Table 4.1. *Example of the FIDES life profile for a helicopter*

However, it is possible to estimate component reliability for a life profile with a variable temperature on the basis of Sedyakin's principle. This is what we succinctly presented in the following section. For more details on the Sedyakin principle, we refer the reader to Nikulin et al. (2007).

4.3. Sedyakin's principle

The Sedyakin (1966) principle proposes the physical principle that assumes that for two identical populations of components operating under the physical contributions X1 and X2 (with X1 ≠ X2), the times t1 and t2 are equivalent if the reliability functions at these two times are equal.

We mathematically obtain that:

$$R_{X1}(t1) = R_{X2}(t2) \quad [4.2]$$

Let us assume that the temperature depends on time in the form of a simple upward step, as shown in Figure 4.1.

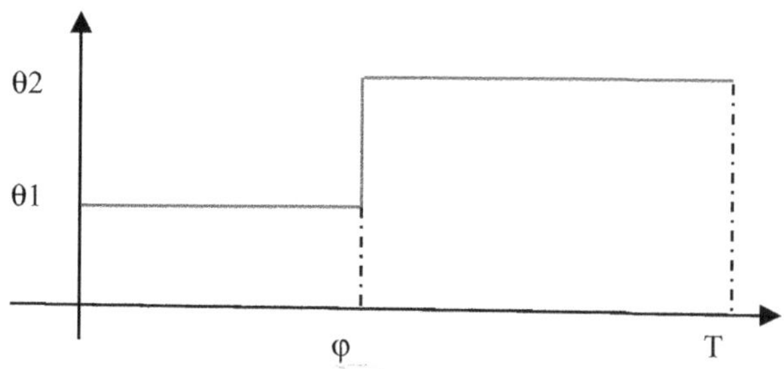

Figure 4.1. *Example of step-up temperature*

To illustrate this principle, we assume that we have a component whose reliability is modeled by an Arrhenius–Weibull law with the following parameters:

$\varphi = 100$; $T = 300$; $\theta 1 = 30°C$; $\theta 2 = 55°C$; $Ea = 0.55$ and $\beta = 2.2$

We obtain the following reliability functions:

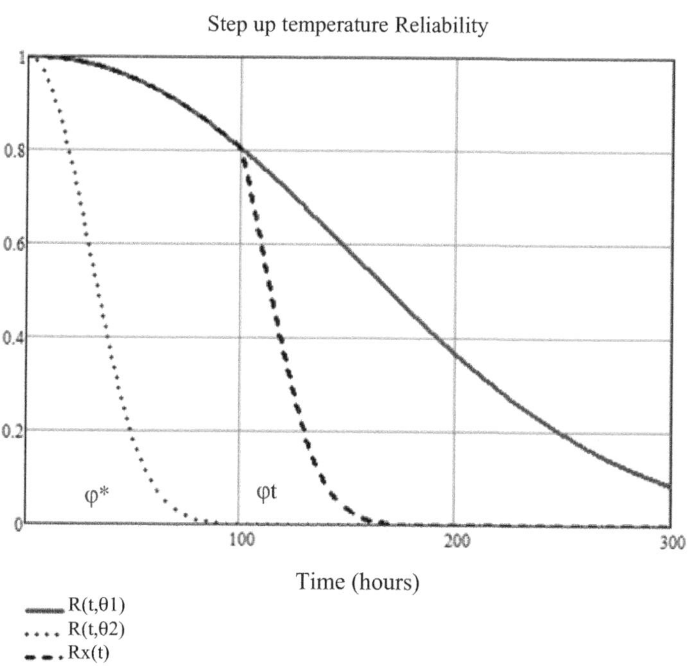

Figure 4.2. *Illustration of the Sedyakin principle for the step-up temperature*

The reliability function of the Arrhenius–Weibull model is given by:

$$R(t,\theta) = \exp\left(-\left(\frac{t}{C.exp\left(\frac{Ea}{Kb.\theta}\right)}\right)^{\beta}\right) \quad [4.3]$$

Equations [4.2] and [4.3] allow us to write the following:

$$\exp\left(-\left(\frac{\varphi}{C.exp\left(\frac{Ea}{Kb.\theta1}\right)}\right)^{\beta}\right) = \exp\left(-\left(\frac{\varphi *}{C.exp\left(\frac{Ea}{Kb.\theta2}\right)}\right)^{\beta}\right)$$

namely:

$$\frac{\varphi}{exp\left(\frac{Ea}{Kb.\theta1}\right)} = \frac{\varphi *}{exp\left(\frac{Ea}{Kb.\theta2}\right)}$$

that is, finally:

$$\varphi *= \frac{\varphi}{exp\left(\frac{Ea}{Kb}.\left(\frac{1}{\theta2} - \frac{1}{\theta1}\right)\right)} \quad [4.4]$$

REMARK.–

In general, the acceleration factor is defined by: if $\theta2 > \theta1$, $AF = exp\left(\frac{Ea}{Kb}.\left(\frac{1}{\theta1} - \frac{1}{\theta2}\right)\right)$ such that the previous equation can also be written as:

$$\varphi *= \frac{\varphi}{AF}$$

4.4. Consequences for reliability estimates

For step-up levels, the temperature is written as (where $\beth$ is the indicator function):

$$\theta(t) = \theta1.\beth_{t\leq\varphi} + \theta2.\beth_{\varphi<t\leq Tfin}$$

Using the Sedyakin principle, the reliability function is then written as:

$$R(t,\theta) = \mathrm{R1}(t,\theta 1).\beth_{t\leq\varphi} + R2\left(t-\varphi+\frac{\varphi}{AF},\theta 2\right).\beth_{\varphi<t\leq Tfin}$$

Although for constant temperatures, the failure mechanism could be modeled by a Weibull distribution, this is no longer the case when the temperature depends on time, as illustrated in Figure 4.3. This result is not specific and would be the same for any other type of distribution (exponential, lognormal, etc.). In reality, not only is this case generally encountered, but we are rather dealing with step-stairs that are periodically renewed. Consider a more realistic example of a system that, every day, undergoes the following thermal evolution:

$$\theta(t) = \theta 1.\beth_{t\leq\varphi 1} + \theta 2.\beth_{\varphi 1<t\leq\varphi 2} + \theta 1.\beth_{\varphi 2<t\leq Tfin}$$

Taking the previous values and setting φ1 = 45; φ2 = 65; Tfin = 100: θ1 = 30°C; θ2 = 45°C; β = 2.2; Ea = 0.55 eV, we obtain:

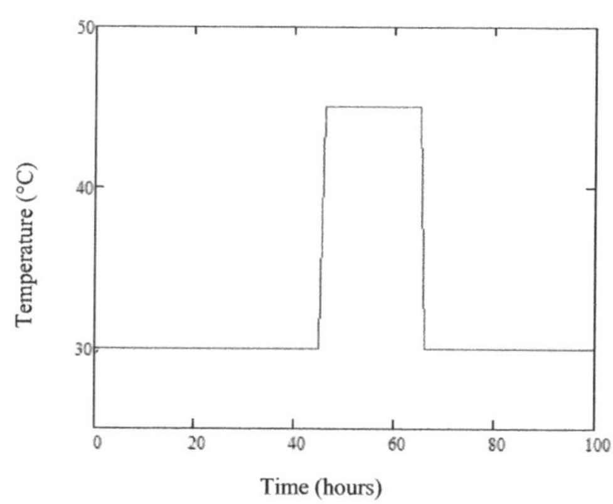

Figure 4.3. *Illustration of a one-period temperature profile*

The corresponding reliability function is then written as:

$$R(t,\theta) = \mathrm{R1}(t,\theta 1).\beth_{t\leq\varphi 1} + R2\left(t-\varphi 1+\frac{\varphi 1}{AF},\theta 2\right).\beth_{\varphi 1<t\leq\varphi 2} + R1\big(t-\varphi 2+AF.\varphi 2-\varphi 1.(AF-1)\big).\beth_{\varphi 2<t\leq Tfin}$$

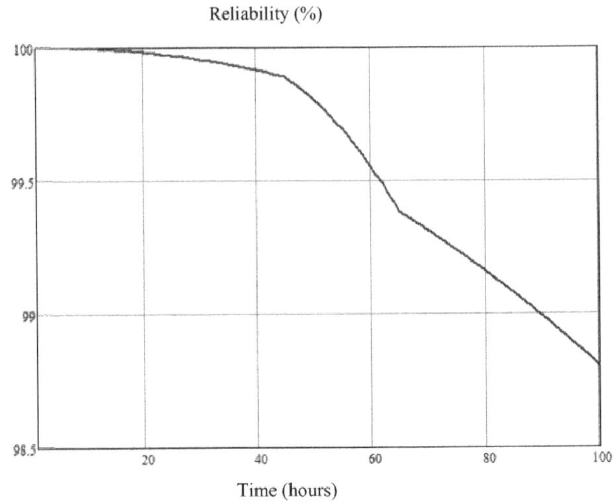

Figure 4.4. *Reliability function of the one-period temperature profile*

Now, consider two periods. We then obtain:

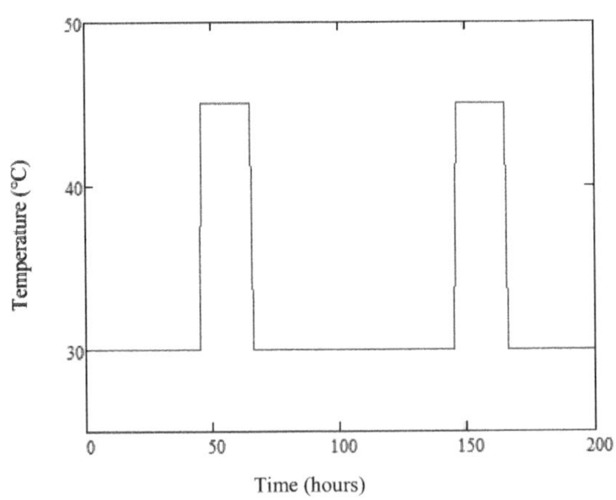

Figure 4.5. *Illustration of a two-period temperature profile*

We then obtain Figure 4.6.

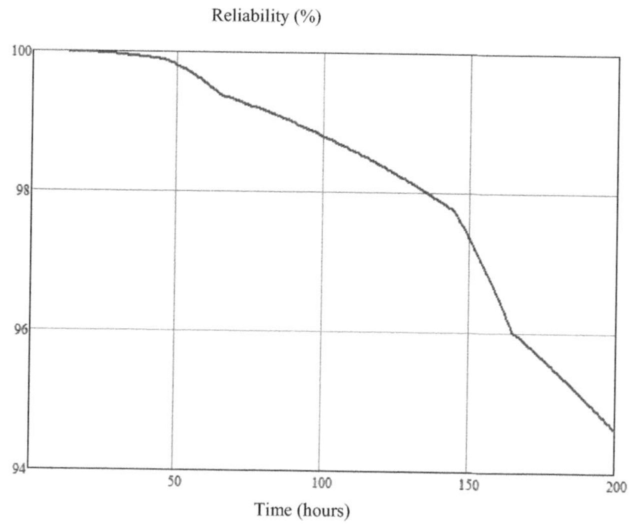

Figure 4.6. *Reliability function of the two-periods temperature profile*

We may then be tempted to concatenate periods of time when temperatures are identical, just as predictive reliability methods typically do. We obtain the following evolution of the temperature:

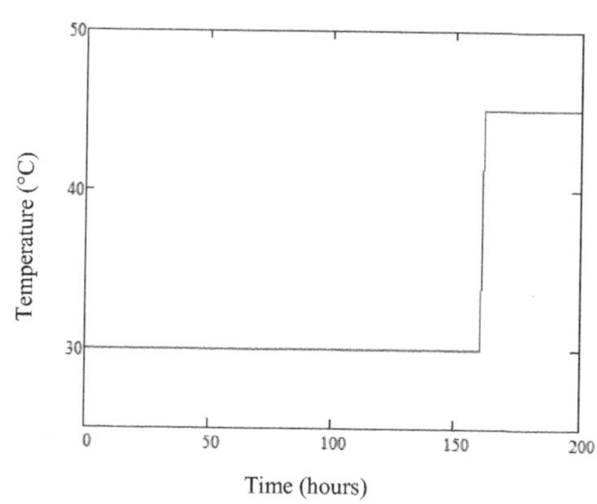

Figure 4.7. *Illustration of a concatenated two-periods temperature profile*

The comparison of the two reliability functions obtained is illustrated in Figure 4.8.

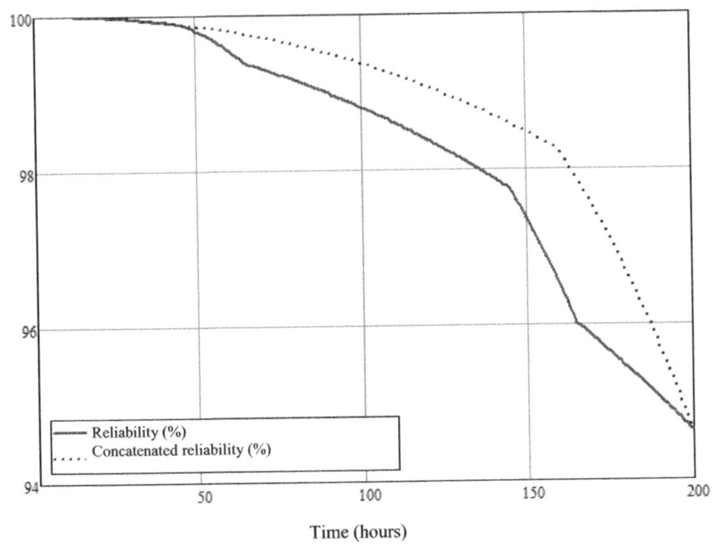

Figure 4.8. *Comparison of the concatenated/unconcatenated temperature profile*

We can see that the two reliability functions do not have the same evolution over time, although the same final values meet each other. Therefore, the method used for predictive reliability estimates, while greatly facilitating calculations, is not rigorously accurate. In practice, the number of periods is actually much larger.

This is the case, for example, for the aeronautics sector but also for railways. In this case, the principle proposed in (Bayle 2019) can be used using a constant equivalent temperature, namely:

$$\theta eq = \frac{{}^{-Ea}/_{Kb}}{ln\left(\frac{1}{T}.\int_0^T exp\left(\frac{-Ea}{Kb.\theta(u)}\right).du\right)} \qquad [4.5]$$

Since we consider a succession of "n" steps, the evolution of the temperature can be written as:

$$\theta(t) = \sum_{i=1}^{n} \theta_i.\beth_{t_{i-1} \leq t < t_i}$$

with $t_0 = 0$ from which if $T = \sum_{i=1}^{n} t_i$:

$$\theta eq = \frac{-Ea/_{Kb}}{ln\left(\frac{1}{T}.\sum_{i=1}^{n}(t_i - t_{i-1}).exp\left(\frac{-Ea}{Kb.\theta_i}\right)\right)} \quad [4.6]$$

Consider the following example: $\beta = 2.2$; Ea = 0.55 eV; $\eta = 5{,}000$ and the following temperature profile.

Time (h)	Temperature (°C)
0–18	15
6–35	45
35–68	30
68–87	50
87–100	20
100–130	15
130–200	60
200–250	37
250–270	12
270–300	25

Table 4.2. *Table of a more realistic temperature profile*

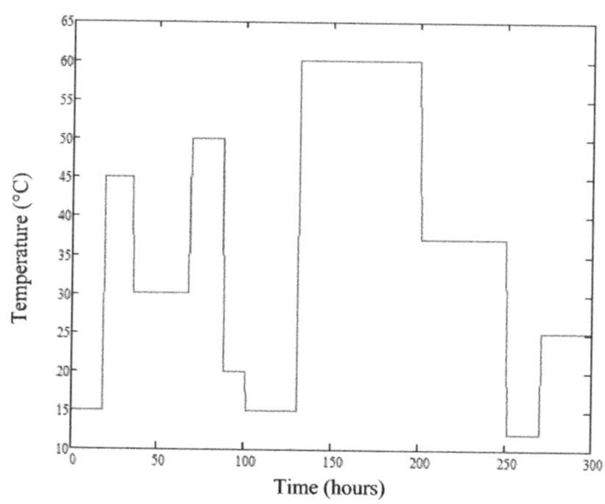

Figure 4.9. *Corresponding temperature time evolution of the more realistic temperature profile*

We obtain Figure 4.9.

The resulting reliability function is given by:

$$R(t,\theta) = \sum_{i=1}^{n} exp\left(-\left(\frac{t - t_{i-1} + \varepsilon_i}{\eta(\theta)}\right)^{\beta}\right) . \beth_{t_{i-1} \leq t < t_i}$$

With the equivalent temperature, we obtain:

$$R(t,\theta eq) = exp\left(-\left(\frac{t}{\eta(\theta eq)}\right)^{\beta}\right)$$

With the numerical data chosen, we obtain:

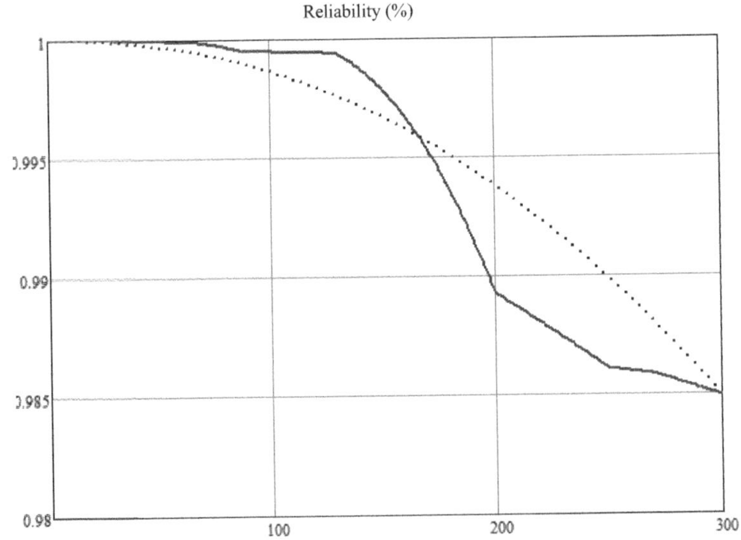

Figure 4.10. *Comparison of the concatenated/unconcatenated temperature profiles*

Equivalence for a single period is incorrect. Now, suppose that the previous temperature profile is repeated ten times. We then obtain Figure 4.11.

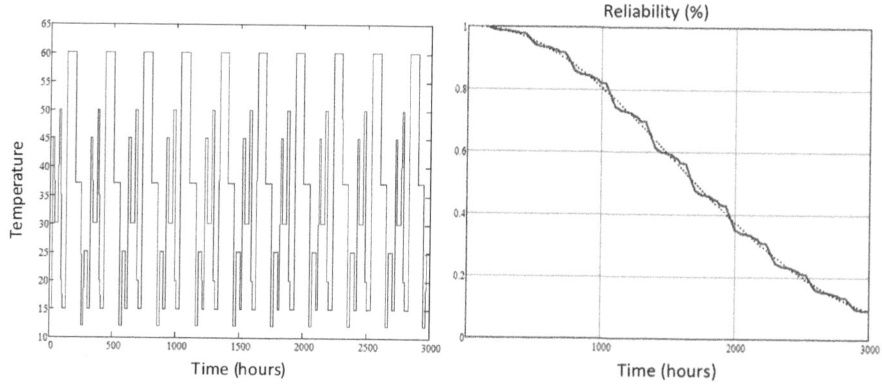

Figure 4.11. *Reliability function for multiple steps*

With only ten periods, the equivalence is quite correct. In practice, the temperature profile is often daily, so that over a period of 20 years, it is repeated 7,300 times. Here, we see why using an equivalent temperature obtained based on the Sedyakin principle can be useful, and the concatenation of time periods with identical temperatures clearly makes sense. This simplifying assumption, which is largely employed in predictive reliability, is then plainly justified.

4.5. Taking the effect of maintenance into consideration

In practice, corrective maintenance is carried out. Since we are at the component level, maintenance will be ideal, and we can then use the theory of renewal processes.

Let us assume that the failure mechanism can be modeled by an Arrhenius-exponential distribution with the following parameters:

$$Ea = 0.4eV;\ \theta 1 = 30°C;\ \theta 2 = 45°C;\ \varphi = 3{,}000;\ Tfin = 10{,}000;\ \lambda 1 = 2.5 \times 10^{-4}$$

Although the Rocof has an explicit solution for a constant temperature, this is no longer the case for temperature steps. Indeed, for upward stair-steps, we have:

$$R(t) = R(t, \theta 1).\beth_{t \leq \varphi} + R\left(t - \varphi + {}^{\varphi}/_{AF}, \theta 2\right).\beth_{t \geq \varphi}$$

Since $f(t) = -\frac{\mathrm{dR(t)}}{\mathrm{dt}}$, we obtain:

$$f(t) = f(t, \theta 1).\beth_{t \leq \varphi} + f\left(t - \varphi + {}^{\varphi}/_{AF}, \theta 2\right).\beth_{t \geq \varphi}$$

Namely,

$$f(t) = \lambda 1.exp(-\lambda 1.t).\beth_{t\leq\varphi} + \lambda 2.exp\left(-\lambda 2.\left(t - \varphi + {}^{\varphi}/_{AF}\right)\right).\beth_{t\geq\varphi}$$

where λ2 = λ1.AF. We obtain the following representation of the probability density.

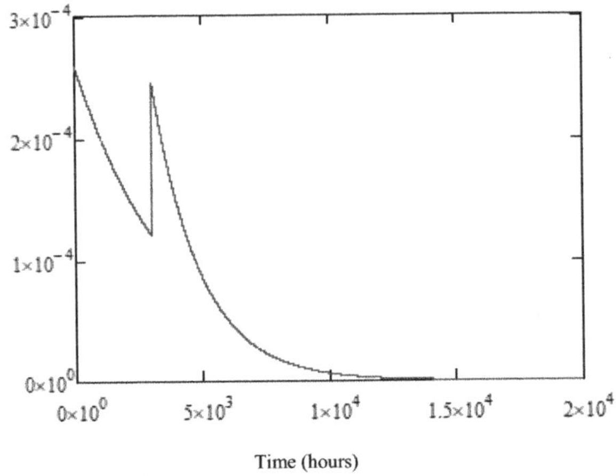

Figure 4.12. *Probability density exponential distribution with ascending temperature stair-steps*

Unlike the reliability function, we have a jump in probability density as the temperature changes. Let us see analytically why.

To this end, we calculate the probability density at time t = φ, that is:

$$f(\varphi) = \lambda 1.exp(-\lambda 1.\varphi) + \lambda 2.exp\left(-\lambda 2.\left(\varphi - \varphi + {}^{\varphi}/_{AF}\right)\right)$$

Namely,

$$f(\varphi) = exp(-\lambda 1.\varphi) + exp\left(-\lambda 2.{}^{\varphi}/_{AF}\right)$$

Now, since θ2 > θ1, we have $\lambda 2 = \lambda 1.AF$ from which:

$$f(\varphi) = \lambda 1.exp(-\lambda 1.\varphi) + \lambda 2.exp(-\lambda 1.\varphi)$$

and thereby finally:

$$f(\varphi) = (\lambda 1 + \lambda 2).exp(-\lambda 1.\varphi)$$

Therefore, with the change in temperature, the density clearly undergoes a discontinuity since there is a shift from $\lambda 1.exp(-\lambda 1.\varphi)$ to $(\lambda 1 + \lambda 2).exp(-\lambda 1.\varphi)$.

The asymptotic value of the Rocof is equal to the inverse of the MTTF given by:

$$MTTF = \int_0^{+\infty} \left(exp(-\lambda 1.t).\beth_{t\leq\varphi} + exp\left(-\lambda 2.\left(t - \varphi + \frac{\varphi}{AF}\right)\right).\beth_{\varphi<t\leq+\infty}\right).dt$$

Namely:

$$MTTF = \int_0^{\varphi} exp(-\lambda 1.t).dt + \int_{\varphi}^{+\infty} exp\left(-\lambda 2.\left(t - \varphi + \frac{\varphi}{AF}\right)\right).dt$$

This is because $\lambda 2 = \lambda 1.\mathrm{AF}$ $MTTF = \frac{1-exp(-\lambda 1.\varphi)}{\lambda 1} + \frac{exp(-\lambda 1.\varphi)}{\lambda 2}$.

Namely, $MTTF = \frac{1}{\lambda 1} + exp(-\lambda 1.\varphi).\left(\frac{1}{\lambda 2} - \frac{1}{\lambda 1}\right)$ from which

$$MTTF = MTTF1 + exp\left(-\frac{\varphi}{MTTF1}\right).(\mathrm{MTTF2} - \mathrm{MTTF1}) \text{ or still}$$

$$MTTF = MTTF1.\left(1 + exp\left(-\frac{\varphi}{MTTF1}\right).\left(\frac{1}{AF} - 1\right)\right) \qquad [4.7]$$

This result is logical because if φ = 0, we obtain a constant value equal to MTTF2, and if φ is infinite, a constant value equal to MTTF1. We then obtain the following Rocof:

Now, let us suppose that the failure mechanism can be modeled by an Arrhenius–Weibull distribution with the following parameters:

Ea = 0.4eV; θ1 = 30°C; θ2 = 45°C; φ = 3,000; Tfin = 10,000; C = 1,162; β = 3.

The probability density for calculating the occurrence rate of Rocof failures is then given by:

$$f(t,\theta) = f(t,\theta 1).\beth_{t\leq\varphi} + f\left(t - \varphi + \frac{\varphi}{AF}, \theta 2\right).\beth_{\varphi<t\leq Tfin}$$

and for an Arrhenius Weibull law:

$$f(t,\theta) = \frac{\beta}{\eta(\theta)} \cdot \left(\frac{t}{\eta(\theta)}\right)^{\beta-1} . exp\left(-\left(\frac{t}{\eta(\theta)}\right)^{\beta}\right)$$

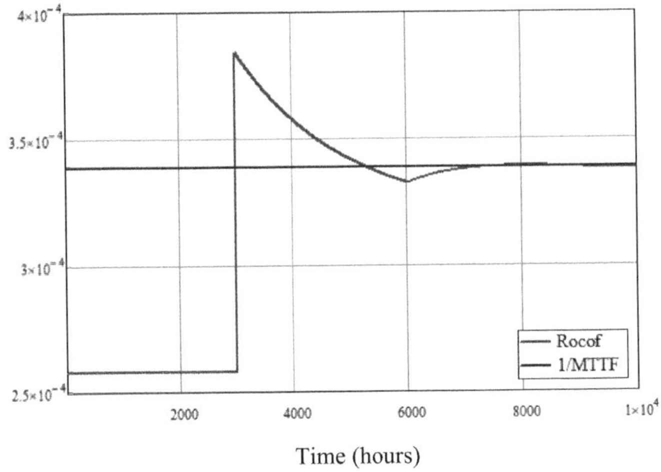

Figure 4.13. *Rocof for an exponential distribution with ascending temperature steps*

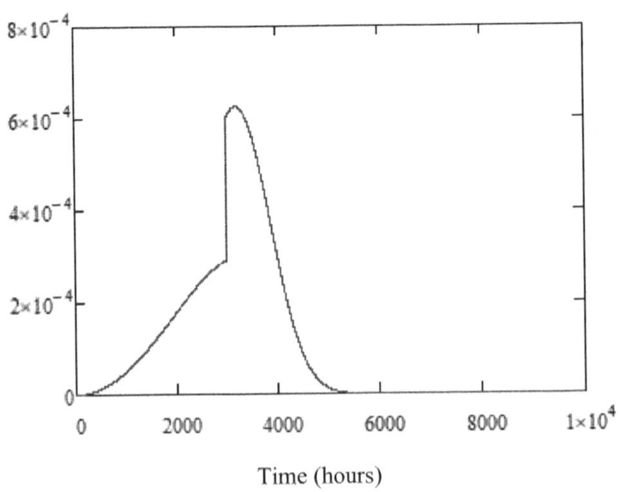

Figure 4.14. *Weibull distribution density with ascending temperature levels*

It is represented in Figure 4.14.

Here, we also observe a discontinuity during the change in temperature. In the same way as for the exponential distribution, it can be written that:

$$f(\varphi) = \frac{\beta}{\eta 1}.\left(\frac{\varphi}{\eta 1}\right)^{\beta-1}.exp\left(-\left(\frac{\varphi}{\eta 1}\right)^{\beta}\right) + \frac{\beta}{\eta 2}.\left(\frac{\varphi/_{AF}}{\eta 2}\right)^{\beta-1}.exp\left(-\left(\frac{\varphi/_{AF}}{\eta 2}\right)^{\beta}\right)$$

This is because η1 = η2. AF.

$$f(\varphi) = \frac{\beta}{\eta 1}.\left(\frac{\varphi}{\eta 1}\right)^{\beta-1}.exp\left(-\left(\frac{\varphi}{\eta 1}\right)^{\beta}\right) + \frac{\beta}{\eta 2}.\left(\frac{\varphi}{\eta 1}\right)^{\beta-1}.exp\left(-\left(\frac{\varphi}{\eta 1}\right)^{\beta}\right) \text{ namely,}$$

$$f(\varphi) = \left(\frac{\beta}{\eta 1}\right).(1 + AF)\left(\frac{\varphi}{\eta 1}\right)^{\beta-1}.exp\left(-\left(\frac{\varphi}{\eta 1}\right)^{\beta}\right)$$

The asymptotic value of Rocof is equal to the inverse of the MTTF. However, for a stair-step, this is given by:

$$MTTF = \int_0^{+\infty}\left(exp\left(-\left(\frac{t}{\eta 1}\right)^{\beta}\right).ב_{t\leq\varphi} + exp\left(-\left(\frac{t-\varphi+\frac{\varphi}{AF}}{\eta 2}\right)^{\beta}\right).ב_{t>\varphi}\right).dt$$

Let $u = \left(\frac{t}{\eta 1}\right)^{\beta}$; thus, $du = \left(\frac{\beta}{\eta 1}\right).\left(\frac{t}{\eta 1}\right)^{\beta-1}.dt$ such that

$$\int_0^{+\infty} exp\left(-\left(\frac{t}{\eta 1}\right)^{\beta}\right).ב_{t\leq\varphi}.dt = \int_0^{\left(\frac{t}{\eta 1}\right)^{\beta}} exp(-u).u^{1/_{\beta}-1}.du$$

We therefore have:

$$\int_0^{+\infty}\left(exp\left(-\left(\frac{t}{\eta 1}\right)^{\beta}\right).ב_{t\leq\varphi}\right).dt$$

$$= \int_0^{+\infty} exp(-u).u^{1/_{\beta}-1}.du - \int_{\left(\frac{t}{\eta 1}\right)^{\beta}}^{+\infty} exp(-u).u^{1/_{\beta}-1}.du$$

Namely:

$$\int_0^{+\infty}\left(exp\left(-\left(\frac{t}{\eta 1}\right)^{\beta}\right).ב_{t\leq\varphi}\right).dt = \Gamma\left(\frac{1}{\beta}\right) - \gamma\left(\frac{1}{\beta},\left(\frac{t}{\eta 1}\right)^{\beta}\right)$$

Let $v = \left(\frac{t-\varphi+\varphi/_{AF}}{\eta 2}\right)^{\beta}$; thus, $dv = \left(\frac{\beta}{\eta 2}\right).\left(\frac{t-\varphi+\varphi/_{AF}}{\eta 2}\right)^{\beta-1}.dt$.

Namely:

$$\int_0^{+\infty} exp\left(-\left(\frac{t-\varphi+\frac{\varphi}{AF}}{\eta 2}\right)^{\beta}\right).\beth_{t>\varphi}.dt = \gamma\left(\frac{1}{\beta},\left(\frac{\frac{\varphi}{AF}}{\eta 2}\right)^{\beta}\right)$$

from which:

$$MTTF = \frac{\eta 1}{\beta}.\left(\Gamma\left(\frac{1}{\beta}\right) - \gamma\left(\frac{1}{\beta},\left(\frac{\varphi}{\eta 1}\right)^{\beta}\right)\right) + \frac{\eta 2}{\beta}.\gamma\left(\frac{1}{\beta},\left(\frac{\varphi}{\eta 2.AF}\right)^{\beta}\right)$$

or still since η1 = η2. AF:

$$MTTF = \frac{\eta 1}{\beta}.\left(\Gamma\left(\frac{1}{\beta}\right) - \gamma\left(\frac{1}{\beta},\left(\frac{\varphi}{\eta 1}\right)^{\beta}\right)\right) + \frac{\eta 1}{\beta.AF}.\gamma\left(\frac{1}{\beta},\left(\frac{\varphi}{\eta 1}\right)^{\beta}\right)$$

or since $\Gamma(1+z) = z.\Gamma(z)$:

$$MTTF = \text{MTTF1} + \frac{\eta 1}{\beta}.\gamma\left(\frac{1}{\beta},\left(\frac{\varphi}{\eta 1}\right)^{\beta}\right).\left(\frac{1}{AF} - 1\right) \quad [4.8]$$

This result makes sense because if φ = 0, then $\gamma\left(\frac{1}{\beta};\left(\frac{\varphi}{\eta 1}\right)^{\beta}\right) = \gamma\left(\frac{1}{\beta};\left(\frac{\varphi}{\eta 2}\right)^{\beta}\right) = \Gamma\left(\frac{1}{\beta}\right)$ from which:

$$MTTF = \text{MTTF1} + \frac{\eta 1}{\beta}.\Gamma\left(\frac{1}{\beta}\right).\left(\frac{1}{AF} - 1\right) = MTTF2$$

If φ = +∞, then $\left(\frac{1}{\beta};\left(\frac{\varphi}{\eta 1}\right)^{\beta}\right) = \gamma\left(\frac{1}{\beta};\left(\frac{\varphi}{\eta 2}\right)^{\beta}\right) = 0$, hence:

$$MTTF = \text{MTTF1}$$

The Rocof is then given by Figure 4.15.

For a parameter with an identical scale, Figure 4.16. shows the evolution of the Rocof for the two distributions.

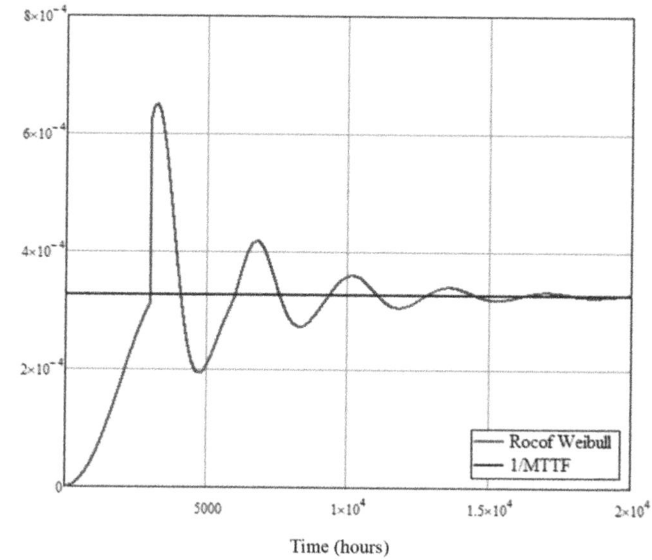

Figure 4.15. *Rocof for a Weibull distribution with a step-up temperature*

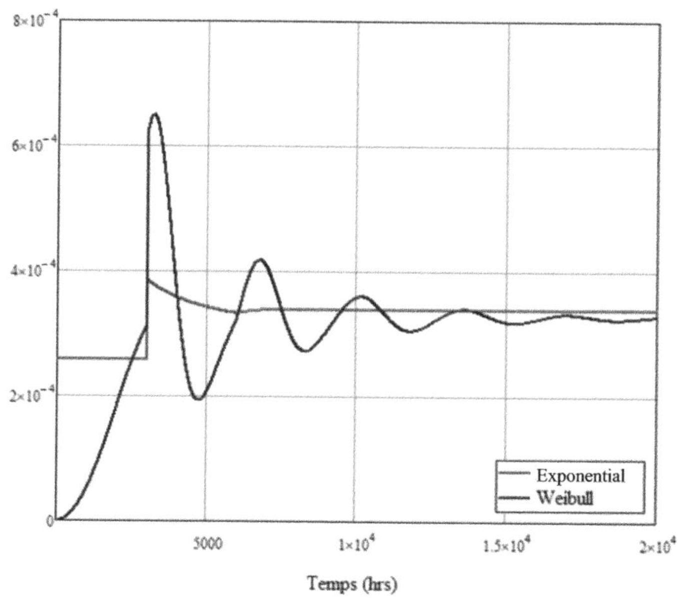

Figure 4.16. *Rocof for the exponential and Weibull distributions for the step-up temperature*

4.6. Summary

– The component reliability depends on the temperature. Modeling following the Arrhenius law is often used in reliability models.

– In the life profiles of manufacturers, temperature levels generally vary from one operational phase to another so that the Arrhenius law is not directly applicable.

– The Sedyakin principle enables the extension of the Arrhenius law for variable temperatures for maintenance-free systems.

– This principle has major consequences for predictive reliability models and accelerated reliability tests.

– For maintenance-enabled systems, there are explicit solutions for the exponential distribution but not for the Weibull distribution.

– We therefore propose, in this case, theoretical calculations as well as Monte Carlo simulations.

5

Aging Tests

We observed in Chapter 3 that the entire team working on the production of a given product tried to avoid youth failures as much as possible, particularly those related to aging, because they will potentially impact the entire range of products that will be delivered to the end user.

Nonetheless, for some industrial applications, the complete eradication of aging during operational activity is not always possible. It is then necessary to build an aging model of the component under consideration to estimate the reliability of the product. This is generally referred to as "aging testing" or "reliability testing".

We also observed that the lifecycle of a product comprises different phases. The aging test is generally performed on a product that is representative of the series. In other words, it is representative of the product definition when it will be delivered to the end user. It is usually implemented after the manufacturing phase, as illustrated in Figure 5.1.

It is clear that an aging test is carried out on a given component and, more particularly, on a specific failure mechanism. Before initiating any procedure related to this type of reliability test, it will be necessary to identify not only the component at risk but also the underlying failure mechanism. Normally, knowledge of the failure mechanism, either by bibliographic analysis or by "business knowledge", will also enable the identification of the type of accelerating physical contribution and therefore the physics law of failure to be employed (Arrhenius, Coffin–Manson, etc.). All of these prerequisites are essential before initiating any such trials. If there is any doubt, it is sometimes even possible to carry out a pretest to verify them.

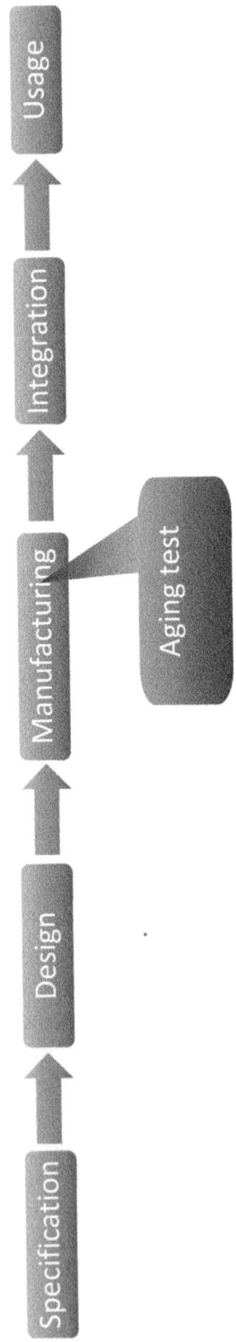

Figure 5.1. *Typical life cycle of systems. For a color version of this figure, see www.iste.co.uk/bayle/montecarlo.zip*

5.1. Accelerated aging test

5.1.1. *Principle*

Suppose that we want to design a product that will operate for 20 years at a temperature of 30°C. The component identified as risky has a heating temperature of 20°C, which means that its internal temperature will be approximately 50°C. To perform an aging test, it is unthinkable to carry out a test under these temperature conditions for 20 years. The duration of the test must be much shorter so that the reliability requirement can be satisfied positively or not in a time compatible with the development of the product.

To overcome this problem, an "accelerated" test must be carried out. To this end, the principle consists of subjecting the component at risk to a level of physical contribution, in our example, a temperature that is higher than that of the operational conditions (in our example, 50°C). Indeed, owing to the Arrhenius equation, the higher the temperature is, the greater the rate of chemical reaction. Therefore, after the industrialization of the transistor, Bondyopadhyay (1998) noted that the higher the temperature is, the faster the components under test fail. In other words, increasing the temperature did not change the failure mechanism but, more importantly, accelerated the time to failure.

In this chapter, we do not cover the practical aspects of a fast-track trial in detail. The choice and adjustment of thermal chambers or shakers, verification of the homogeneity of the levels of the physical contributions to which the components are subjected, assembly diagram of the components under test, test and monitoring methods will deliberately be not taken into consideration because they are independent of the Monte Carlo simulations. Nonetheless, it is clear that all of this is also very important for a successful accelerated trial. For the record, we participated in an accelerated test on DC/DC voltage converters that had a "Boolean" output, which signals their proper functioning and on which a failure LED-based visualization had been implemented. Some of them switched off, meaning failure in the converter, after a few hundred temperature cycles. The converters worked very well, but the LEDs were faulty because they had been placed in the thermal chamber and therefore were also subjected to severe thermal cycles.

5.1.2. *Modeling*

We also observed that a Weibull distribution was generally employed to model the reliability of the three different types of failure and was well adapted, especially for aging.

As a reminder, this can be defined by its failure rate given by:

$$\lambda(t) = \frac{\beta}{\eta}.\left(\frac{t}{\eta}\right)^{\beta-1}$$

We previously reported that increasing the level of physical contribution is equivalent to increasing the speed of time. Therefore, we propose the use of the AFT model, which, for a Weibull distribution and a given physical contribution X, results in:

$$\lambda(t,X) = \frac{\beta}{\eta(X)}.\left(\frac{t}{\eta(X)}\right)^{\beta-1}$$

In the remainder of this chapter, we focus on temperature because it is often this physical contribution that appears in failure mechanisms, but the following is quite applicable to other physical contributions where only the physics law of failure changes. We thus propose the use of the Arrhenius law for a constant temperature such that we obtain:

$$\lambda(t,\theta) = \frac{\beta}{\mathrm{C.exp}\left(\frac{E_a}{K_b.\theta}\right)}.\left(\frac{t}{\mathrm{C.exp}\left(\frac{E_a}{K_b.\theta}\right)}\right)^{\beta-1}$$

The problem that arises is now that the parameters C and E_a are unknowns, which implies that at least two different temperatures will be needed to estimate them. The choice of these two (or more) temperatures will then largely determine either the duration of the test or the number of failures observed but also the quality of the estimation of these parameters.

5.2. Aging test design

To achieve the best design for this type of test, it is essential to know the input data, the parameters that can be manipulated and the industrial constraints of the product.

5.2.1. *Input data*

First, a level of trust has to be defined unless it is already established in the customer reliability requirement. The reliability requirement will be estimated from the reliability allowance made at the product level (Bayle 2021). It will also be essential to have a life profile that is as faithful as possible to operational conditions. Finally, for the "temperature" contribution, it will be necessary to consider the self-heating process of the tested component if there is one. This can be determined either by calculation, preferably because it is much more realistic, or by thermal simulation based on finite elements or by measurement (infrared camera, for example).

5.2.2. *Adjustment parameters*

The possible adjustment parameters are generally as follows:

– the number of components under test;

– the maximum temperature of the test;

– the minimum temperature of the test;

– the number of temperatures in the test;

– the type of test (constant levels, step levels, etc.);

– the duration of the test;

– the testing strategy (sequential, parallel or mixed).

5.2.3. *Product constraints*

The product constraints can be the following:

– The maximal temperature of the component guaranteed by the manufacturer. This is given in the technical documentation of the component (datasheet). It is strongly advised not to exceed this maximal value because the component may no longer hold its full performance or will be the site of a failure mechanism different from the one sought after, especially under operational conditions.

– The number of test methods because this parameter is highly important for the test strategy. Indeed, if only one thermal chamber is available for the test, it will necessarily be sequential.

– The type of monitoring method. To model the failure mechanism, the failure times of all of the components under test must be measured as accurately as possible. However, this is not always possible, especially for verifying certain characteristics of the component, and sometimes the test needs to be stopped to test the component on a specific device.

– The maximal number of components per test process. For the temperature, it will be necessary to have a thermal chamber in which the volume is necessarily restricted. Therefore, depending on the volume of the component under test, the number of components cannot exceed a maximum value.

– The duration of the test as this can be imposed by the project.

– The cost of the test because it may be imposed by the project.

5.2.4. *Impact of the shape parameter β of the Weibull distribution*

We observed in Chapter 2 that the shape parameter β had a physical meaning because it represented the aging kinetics of the failure mechanism under study. Therefore, the greater it is, the faster the failures occur as soon as one of them is observed. This fact can then have a major impact on the number of components being tested. Indeed, the greater the shape parameter is, the more the components tend to behave in the same way, that is, they fail in a very short period of time. If this is the case, then it is not necessarily useful to have a large number of components, which will substantially reduce costs and testing time.

This parameter is of course not known before the test, but a bibliographic analysis or a pretest can give a good idea of its value.

5.2.5. *Choosing the minimal and maximal temperatures*

These parameters can be adjusted, but they are also subject to constraints. For the maximum temperature, the maximum value guaranteed by the component manufacturer must absolutely not be exceeded. The latter therefore sets the maximum value of the test. However, it is important to properly take into account self-heating in the component under operational conditions.

The minimum temperature is also quite simple since it is absolutely essential to speed up the time and therefore to have a temperature higher than or equal to that seen by the component under operational conditions. From these two pieces of information, we thus have a range of possible temperature values.

Which of these is the most significant lowest temperature value?

To illustrate our point, let us use the following example. Consider a failure mechanism of a temperature-accelerated component that can be modeled by a Weibull–Arrhenius distribution with the following parameters:

– Ea = 0.7 eV;

– $\beta = 3$;

– $\eta = 1{,}000$ h @ 125°C;

– guaranteed maximal component temperature of 125°C;

– operating temperature of 50°C.

5.3. Sequential test at two constant temperatures

In this chapter, we consider an accelerated test with two temperature levels performed sequentially, as shown in Figure 5.2.

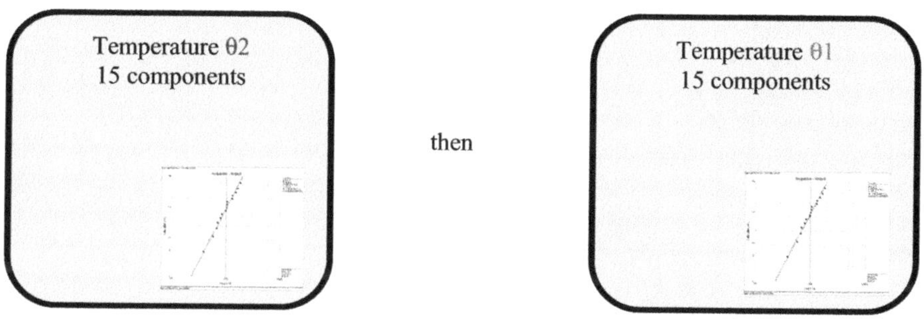

Figure 5.2. *Illustration of sequential accelerated tests for temperature stress. For a color version of this figure, see www.iste.co.uk/bayle/montecarlo.zip*

We already saw in Chapter 2 how important is the number of components during an accelerated test. Suppose that the thermal chamber should not hold more than fifteen components, whether to avoid cluttering the layout or due to considerations of cost or related to testing and monitoring mechanisms.

The highest temperature T2 should be the best starting point to obtain a reasonably short testing time and therefore failure information as quickly as possible. Therefore, we choose T2 = 120°C.

On the basis of the principle of inversion of the probability of failure, fifteen times are thus simulated until failure according to the law taken as an assumption.

Presupposing that all components being tested have failed, a Weibull distribution model based on the maximum likelihood method is shown in Figure 5.3.

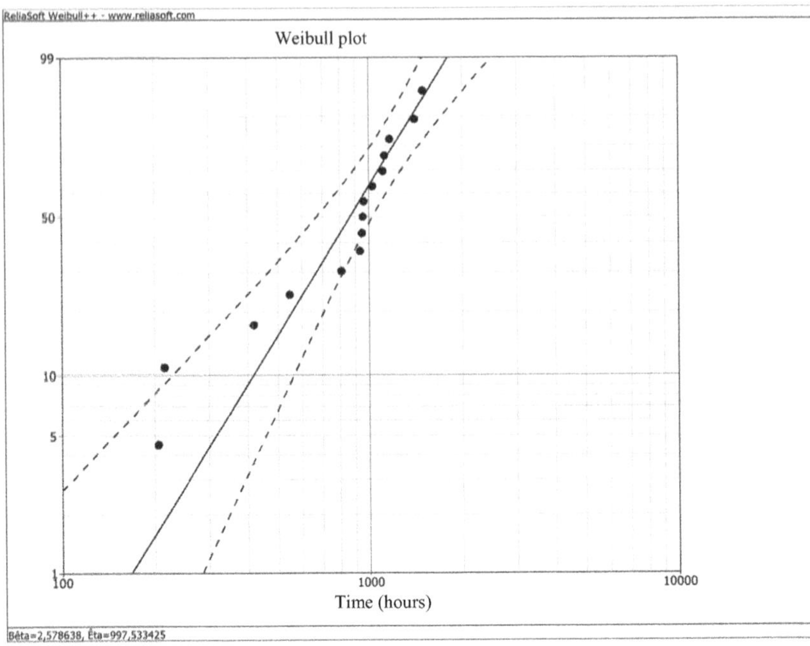

Figure 5.3. *Corresponding Weibull probability function. For a color version of this figure, see www.iste.co.uk/bayle/montecarlo.zip*

The expertise of the faulty components shows that we have indeed observed the suspected failure mechanism before the test. The question that arises is *"what temperature θ1 should I choose?"*.

To answer this question, we propose the use of Monte Carlo simulations to determine the influence of this parameter when all of the other parameters are known.

The parameter of interest is, of course, the activation energy of the failure mechanism. This is of course not known because if it did, the aging test would have no use. However, a literature review or a pretest can provide an estimate.

The criterion for determining the influence of temperature θ1 and the model dispersion that can be visualized is the probability of failure of the component under operational conditions. We therefore propose measuring this dispersion for the following temperatures: θ1 = 110, 100, 80 and 60°C. We obtain the following figures for 1,000 simulations.

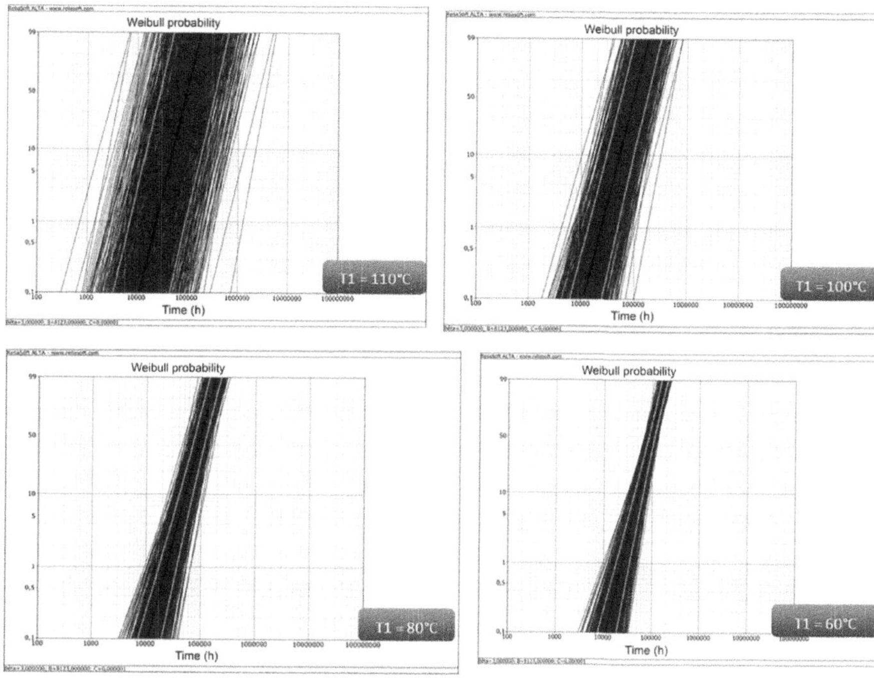

Figure 5.4. *Variability Monte Carlo simulations for different temperature levels. For a color version of this figure, see www.iste.co.uk/bayle/montecarlo.zip*

Visually, on the four previous curves, the lower the minimal temperature is, the less dispersed the reliability data are. This is because the acceleration factor, given by $AF = \exp\left(\frac{Ea}{Kb}\cdot\left(\frac{1}{\theta 1} - \frac{1}{\theta 2}\right)\right)$, is larger for lower θ1. The downside is that failure times are much greater since the times to failure for temperature θ2 are multiplied by this acceleration factor for temperature θ1.

Let us now look at the effect of the activation energy by redoing these simulations but for Ea = 0.4 eV. More precisely, we estimate data dispersion on the basis of the coefficient of variation (CV) as the ratio between the standard deviation and the mean. We obtain the results shown in Figure 5.5.

As we have previously observed, the dispersion is lower when the temperature θ1 decreases. On the other hand, it is greater when the activation energy is lower because, for the same temperature θ1, the acceleration factor is lower for a lower activation energy. If we set a maximum coefficient of variation of 10%, we obtain θ1 = 100°C and AF ~ 13.5 for Ea = 0.7 eV, while θ1 = 80°C and AF ~ 4.4 for

Ea = 0.4 eV. Since we have fixed the number of components being tested, the duration of the test cannot be known in advance since it is random.

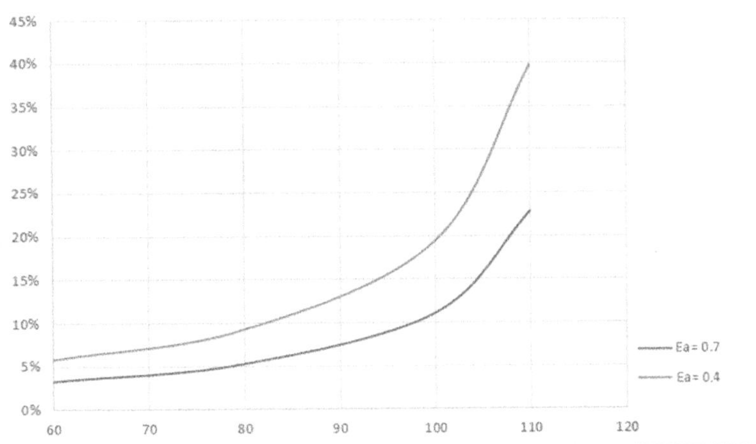

Figure 5.5. *Coefficient of variation of Ea according to temperature. For a color version of this figure, see www.iste.co.uk/bayle/montecarlo.zip*

5.4. Constant-level parallel testing

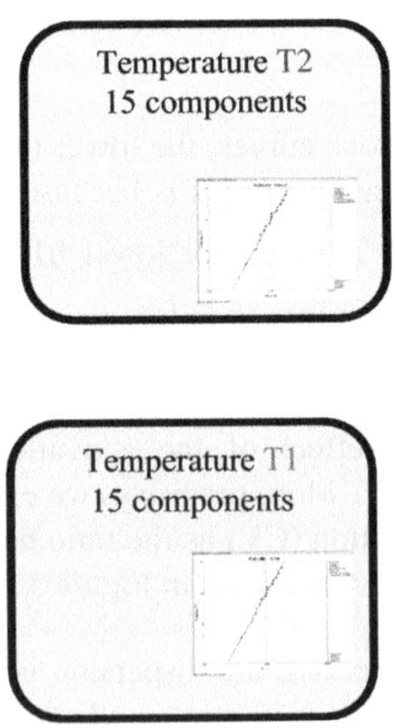

Figure 5.6. *Illustration of parallel accelerated tests for temperature stress. For a color version of this figure, see www.iste.co.uk/bayle/montecarlo.zip*

In this chapter, we consider an accelerated test with two temperature levels performed in parallel, as shown in Figure 5.6.

This means that we have two thermal chambers and that the tests at both temperatures are carried out at the same time. The one with the highest temperature will finish earlier. The choice of temperatures is identical to that of the sequential tests.

5.5. Constant-level mixed testing

This is actually a mix of serial and parallel test conditions. Depending on the number of conditions to be tested, there are more or fewer possible tests. At least three test conditions are needed, which leads, in the most common case of three test conditions, to the following scheme:

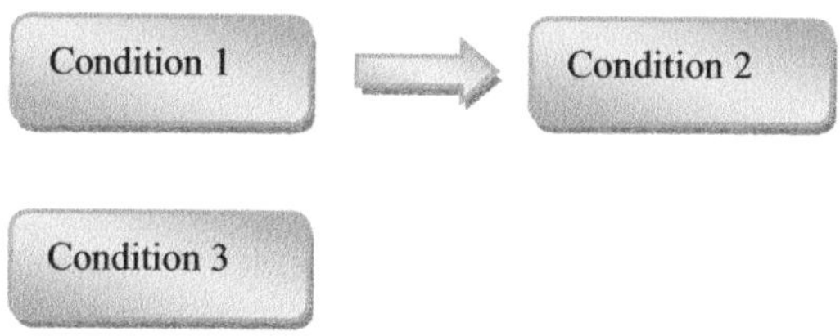

Figure 5.7. *Illustration of mixed accelerated tests for temperature stress*

This case is useful, especially when we do not obtain a failure in Condition 2, when the literature review indicated that this should have been the case. This finding may be because the estimate obtained from the acceleration factor is incorrect, either because it is lower than expected or because the expected failure mechanism is not activated for this level of physical constraint.

Therefore, to clarify any doubt, if possible, an intermediate level of physical contribution (Condition 3) is carried out between Conditions 1 and 2 with a second test method.

5.6. Summary

– In some industrial contexts, reliability testing on a specific component exhibiting an aging failure mechanism will necessarily have to be performed.

– Depending on the context, a certain amount of input data (confidence level, maximum component temperature, etc.) and constraints (testing, monitoring methodology, test costs, etc.) are then evaluated.

– This is followed by a certain number of parameters for test adjustment (test strategy, number of components under test, choice of temperature levels, etc.).

– On the basis of a bibliographic analysis of the failure physics of the component, a choice of test parameters (reliability model, minimal and maximal temperatures, number of components, etc.) is made.

– Monte Carlo simulations, which are performed with the failure probability inversion method, make it possible to evaluate the statistical properties of the parameters of the chosen model and thus optimize the reliability test.

6

Application of the Noncentral Beta Distribution

6.1. Context

It may be that, for some particular problems, classical methods and reliability analysis models are not suitable or do not solve the problem at hand satisfactorily. It is then necessary to resort to other probability distributions to eventually find a definitive solution to the problem under consideration. In general, in terms of reliability, the domain of definition of the random variable of interest is the set of positive real numbers. This observation somewhat restricts the continuous probability distributions to the conventional exponential, Weibull or lognormal models. The normal distribution is therefore rarely employed because of its symmetry with respect to the mean value and the possibility of taking into account negative values of the variable of interest, especially for the times until failure.

In our industrial experience, we encountered a case in which all these continuous probability distributions did not yield satisfactory results. This was specifically the case with a component in which a manufacturing defect caused a short circuit between the signal and the electrical ground during its operational use. After X-ray observation of a sample of components, although not all of them were short-circuited, it was observed that a whole batch of this type of component had the same type of defect to a greater or lesser extent.

Knowing that this failure mode was critical for the application being considered, the following question emerged: *"what is the probability of having a short circuit for all the components that are in operational activity?"*.

As mentioned earlier, none of the conventional probability distributions gave a satisfactory answer, and we then considered other distributions. In the measurements performed to estimate the distance between the signal and the ground, we note a "large distribution tail" in the observed distribution, which is the main obstacle for "classical" probability distributions, as illustrated below, with a representation in the form of a box plot.

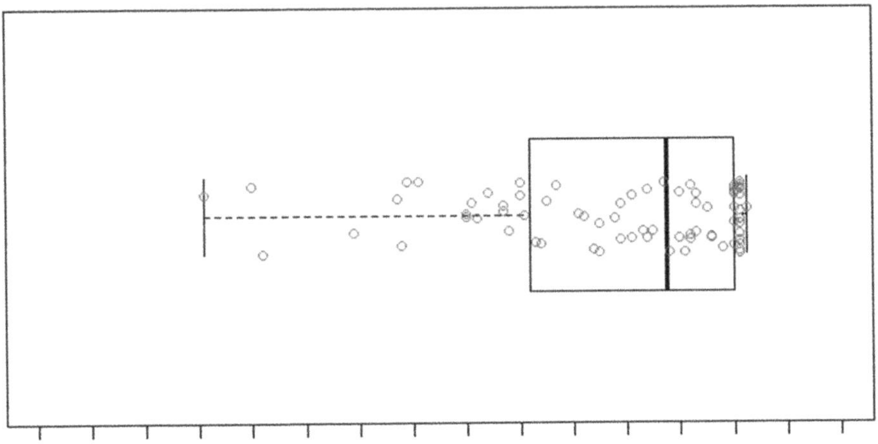

Figure 6.1. *Measurement points between the signal and ground. For a color version of this figure, see www.iste.co.uk/bayle/montecarlo.zip*

In addition, the range of possible values was finite, namely, between 0 and a maximal value. By analyzing the continuous probability distributions that can potentially be utilized, we arrived at the beta probability distribution. It admits a wide variety of shapes according to its two parameters α and β. It indeed has bounded support but is defined only on the interval [0, 1], which at first glance is not adequate for our problem.

However, by normalizing the distance between the signal and the ground by the distance between them, we obtain the interval [0, 1], thus validating the use of the beta probability distribution.

NOTE.– The probability distribution should not be mistaken for the real function.

For the normalized distance, because of the large tail toward values close to 1, this probability law, although it yields better results than the conventional distributions, was not completely satisfactory because of the criticality of the event generated by the short circuit. We then propose the noncentral beta probability distribution, which has greater potential to solve our problem.

6.2. The "noncentral beta" probability distribution

It is defined by its probability density defined by:

$$f(t,\alpha,\beta,\lambda) = \sum_{j=0}^{+\infty}\left[\frac{1}{j!}.\left(\frac{\lambda}{2}\right)^{j}.exp\left(-\frac{\lambda}{2}\right).\frac{t^{\alpha+j-1}.(1-t)^{\beta-1}}{B(\alpha+j.\beta)}\right].\gimel_{x\in[0,1]} \qquad [6.1]$$

where:

– B is the beta function defined by $B(a,b) = \int_0^1 t^{a-1}.(1-t)^{b-1}.dt$;

– α and β shape parameters;

– λ the noncentrality parameter.

Figure 6.2 gives some possible patterns of this density according to these three parameters.

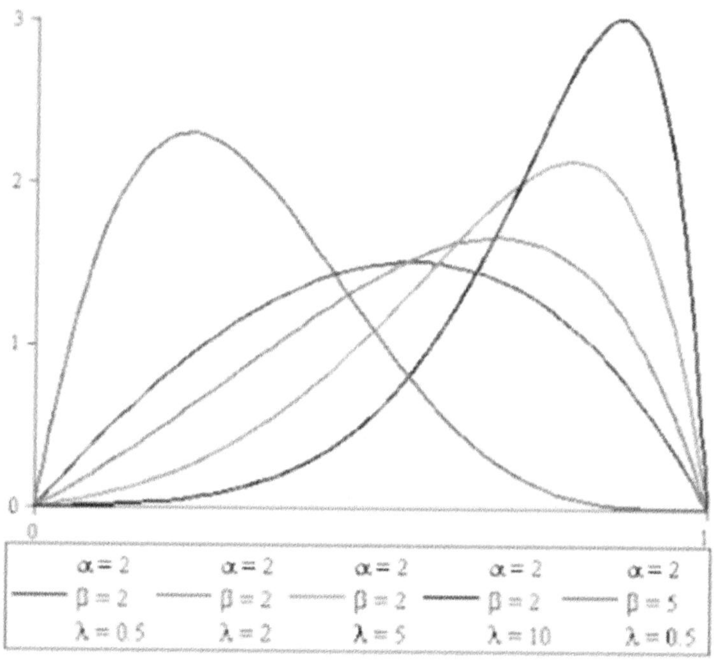

Figure 6.2. *Probability density of the noncentral beta distribution. For a color version of this figure, see www.iste.co.uk/bayle/montecarlo.zip*

The cumulative distribution function of the noncentral beta distribution is given by:

$$F(t,\alpha,\beta,\lambda) = \sum_{j=0}^{+\infty}\left[\frac{1}{j!}.\left(\frac{\lambda}{2}\right)^{j}.exp\left(-\frac{\lambda}{2}\right).I_t(\alpha+j,\beta)\right] \quad [6.2]$$

where the regularized incomplete beta function is defined by:

$$I_t = \frac{\int_0^t x^{a-1}.(1-x)^{b-1}.dx}{\int_0^1 x^{a-1}.(1-x)^{b-1}.dx}$$

The probability of failure of this distribution is complex, and the inversion method described in the previous chapters cannot be used for generating random values according to this probability distribution. We then propose the use of the rejection method.

6.3. Measurement modeling

Now, the noncentral beta distribution parameters need to be estimated relative to the data in Figure 6.1. We propose the use of the maximum likelihood method. The likelihood function, which describes the measures x_i of a statistical distribution as a function of the parameters α, β and λ, is defined by:

$$L(\alpha,\beta,\lambda) = \prod_{i=1}^{n} f(x_i,\alpha,\beta,\lambda)$$

where n is the number of measurements

It is preferable to use the log-likelihood from which:

$$\Lambda(\alpha,\beta,\lambda) = ln\big(L(\alpha,\beta,\lambda)\big) = \sum_{i=1}^{n} ln\big(f(x_i,\alpha,\beta,\lambda)\big) \quad [6.3]$$

The estimates of the parameters α and βλ are then obtained by solving the following system of equations:

$$\frac{\partial\Lambda(\alpha,\beta,\lambda)}{\partial\alpha} = 0$$

$$\frac{\partial\Lambda(\alpha,\beta,\lambda)}{\partial\beta} = 0$$

$$\frac{\partial\Lambda(\alpha,\beta,\lambda)}{\partial\lambda} = 0$$

Specifically, via equations [6.1] and [6.3], we get:

$$\frac{\partial}{\partial\alpha}\sum_{i=1}^{n} ln\left(\sum_{j=0}^{+\infty}\left[\frac{1}{j!}\cdot\left(\frac{\lambda}{2}\right)^{j}.exp\left(-\frac{\lambda}{2}\right).\frac{t^{\alpha+j-1}.(1-t)^{\beta-1}}{B(\alpha+j.\beta)}\right].\daleth_{x\in[0,1]}\right)=0 \qquad [6.4]$$

$$\frac{\partial}{\partial\beta}\sum_{i=1}^{n} ln\left(\sum_{j=0}^{+\infty}\left[\frac{1}{j!}\cdot\left(\frac{\lambda}{2}\right)^{j}.exp\left(-\frac{\lambda}{2}\right).\frac{t^{\alpha+j-1}.(1-t)^{\beta-1}}{B(\alpha+j.\beta)}\right].\daleth_{x\in[0,1]}\right)=0$$

$$\frac{\partial}{\partial\lambda\sum_{i=1}^{n} ln\left(\sum_{j=0}^{+\infty}\left[\frac{1}{j!}\cdot\left(\frac{\lambda}{2}\right)^{j}.exp\left(-\frac{\lambda}{2}\right).\frac{t^{\alpha+j-1}.(1-t)^{\beta-1}}{B(\alpha+j.\beta)}\right].\daleth_{x\in[0,1]}\right)}=0$$

This system of equations has no explicit solutions, and a numerical solution method is needed. We then obtain the following representation.

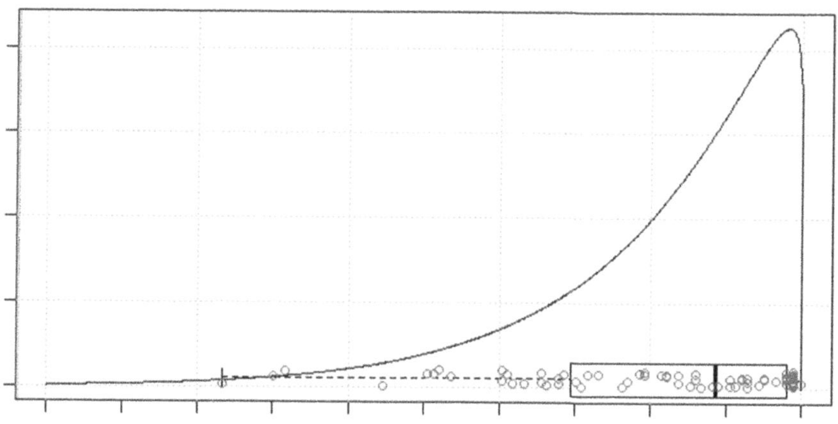

Figure 6.3. *Probability density function of the decentered beta distribution obtained from likelihood method. For a color version of this figure, see www.iste.co.uk/bayle/montecarlo.zip*

From this model, it is then possible to answer the question in section 6.1, "*what is the probability of having a short circuit for all of the components that are in operational activity?*".

Of course, we have only one experiment, and it would not be wise to conclude that estimating the component probability of failure represents the true value with a single dataset. It is therefore very important, from a level of confidence that we establish, to find a confidence interval for this probability. For Chapter 2, we propose the use of the inversion method but given the expression of the probability of the noncentral beta distribution given by equation [6.2], this is not possible, and we propose the use of the rejection method.

6.4. Rejection method

In some cases, the inversion method cannot be utilized because an analytic expression of the reciprocal function of the probability of failure cannot be found. This makes it possible to generate a random variable X on the basis of the acceptance (or rejection, hence its name) of the generation of another variable Y encompassing the desired distribution. The algorithm is given below. It generates a random variable X of probability density f, using the probability density g of another distribution that we know how to generate.

The choice of c conditions the number of generations performed, which is proportional to $\frac{1}{c}$. The strong point of this simulation method is its simplicity of implementation. The weak point is the need to reject a number of draws proportional to $\frac{1}{c}$, but iterative methods can be used to optimize the value of c. To better understand the method, consider the following example with the noncentral beta distribution of parameters $\alpha = 2$, $\beta = 2$ and $\lambda = 10$. The corresponding probability density is obtained from equation [6.1].

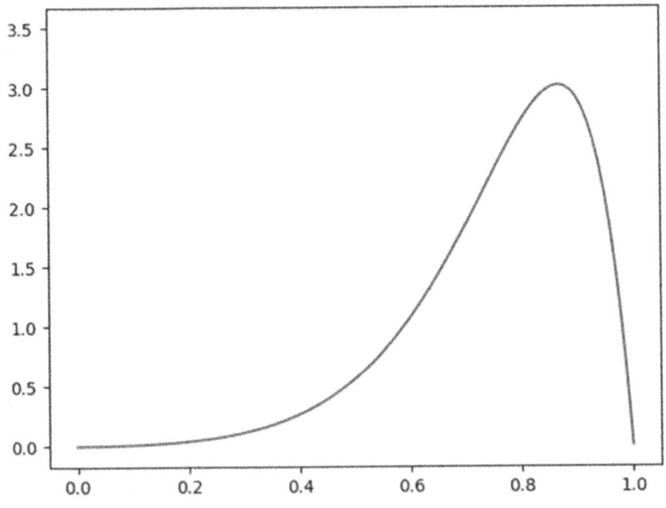

Figure 6.4. *Probability density function of the chosen example*

Let the function g follows a uniform distribution of support [0; 1]. The value of c is chosen so that it has c.g ≥ f, namely, for example, c = 3.4.

First iteration

Let Y1 be a random value according to g: Y1 = 0.8143.

Let U1 be a random value generated according to c.g(Y1): U1 = 3.4 × 0.4016 = 1.4056.

The generated point of coordinates (0.8143; 1.4056) is indicated in Figure 6.5 by a green cross.

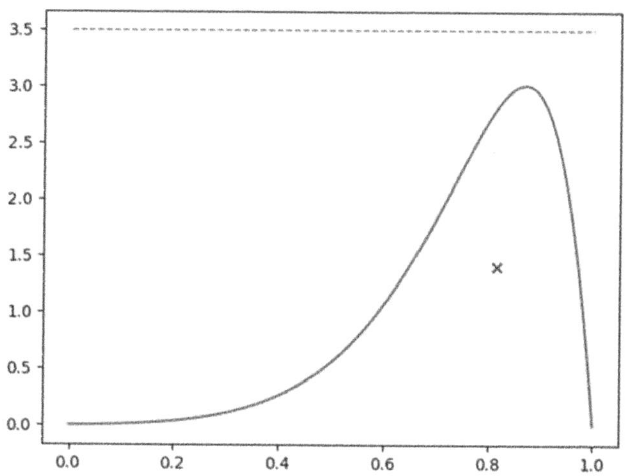

Figure 6.5. *Probability density function with one iteration*

The value of f for 0.8143 is obtained from equation [6.1]. Since U1 < f, X1 = 0.8143. This value X1 is accepted as a simulation.

Second iteration

Let Y2 be a random value according to g: Y2 = 0.6612.

Let U2 be a random value generated according to c.g(Y2): U2 = 3.5 × 0.9709 = 3.3984. Since U2 > f, U2 is rejected.

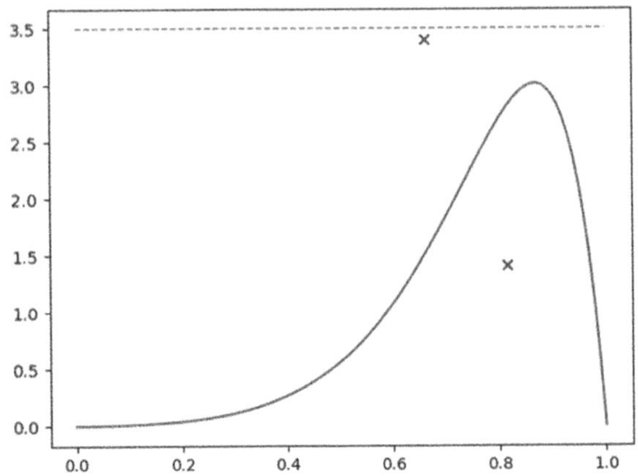

Figure 6.6. *Probability density function with two iterations. For a color version of this figure, see www.iste.co.uk/bayle/montecarlo.zip*

The algorithm is iterated until a value for X2 is obtained. In fact, while the condition U2 < f, we try another random value for Y2 until the above condition is fulfilled. In the example above, to obtain one point, it was necessary to run five generations with four rejections and one acceptance.

The algorithm is restarted to obtain 100 values, and we obtain the results in Figure 6.7.

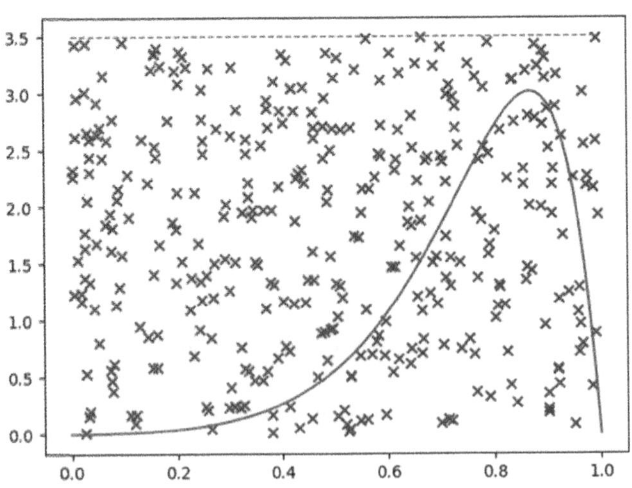

Figure 6.7. *Probability density function of a large iteration simulation. For a color version of this figure, see www.iste.co.uk/bayle/montecarlo.zip*

We had to run 357 generations to obtain our 100 points. This gives an acceptance ratio of 100/357 = 0.28, which is close to 1/c = 1/3.5 = 0.29. Therefore, all of the ordinates of the green points are taken into account for the simulation.

OBSERVATIONS.–

– The choice of value for the parameter "c" is important. Too large a value will generate many generations, whereas too low a value may imply a poor simulation of the curve sought after.

– The choice of support "g" is also important. Indeed, support that is very far from the desired curve will induce a large number of generations. This is what we see in the example above with the choice of the uniform distribution.

We therefore obtain the computation algorithm shown in Figure 6.8.

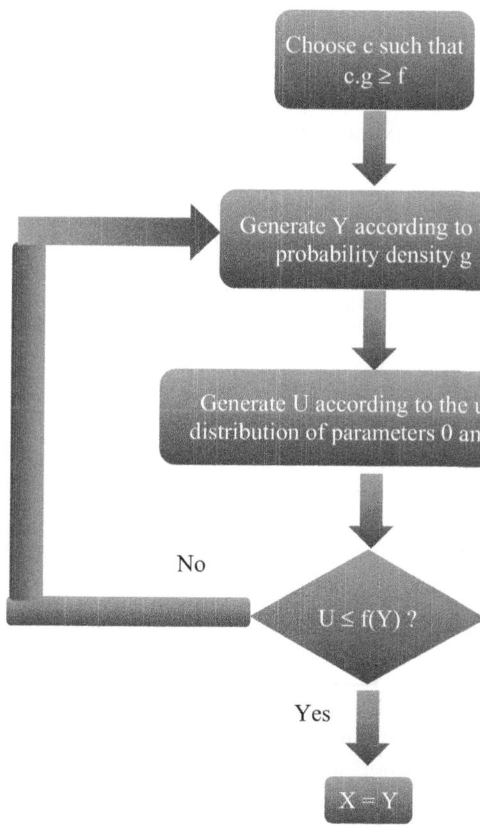

Figure 6.8. *Rejection method algorithm*

6.5. Confidence interval for noncentral beta distribution

On the basis of Monte Carlo simulations using the rejection method, it is possible to obtain a confidence interval for the probability of a dreaded event (short circuit between the signal and ground). The algorithm shown in Figure 6.9 is then obtained.

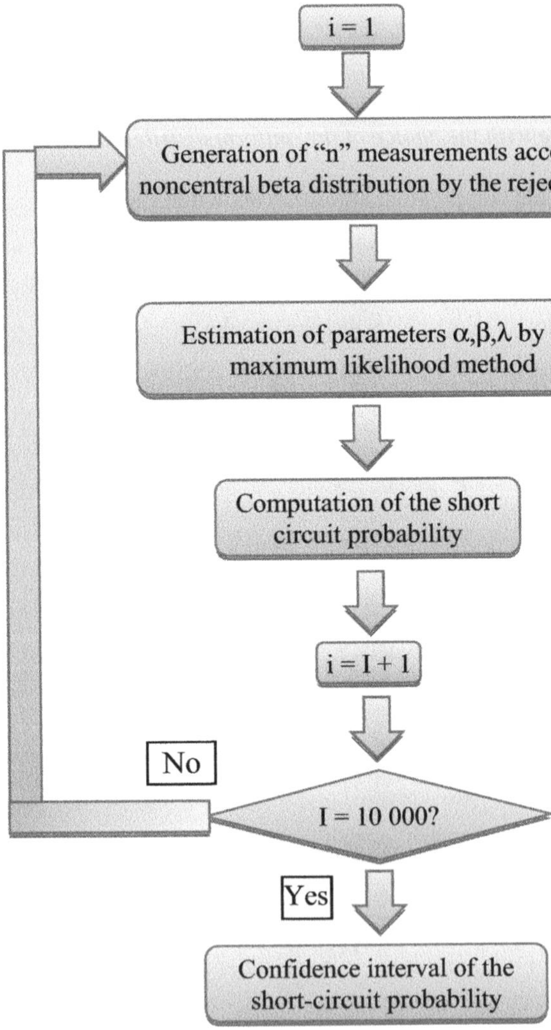

Figure 6.9. *Algorithm for the generation of a confidence interval*

If we choose a 90% confidence level, we obtain the results in Figure 6.10.

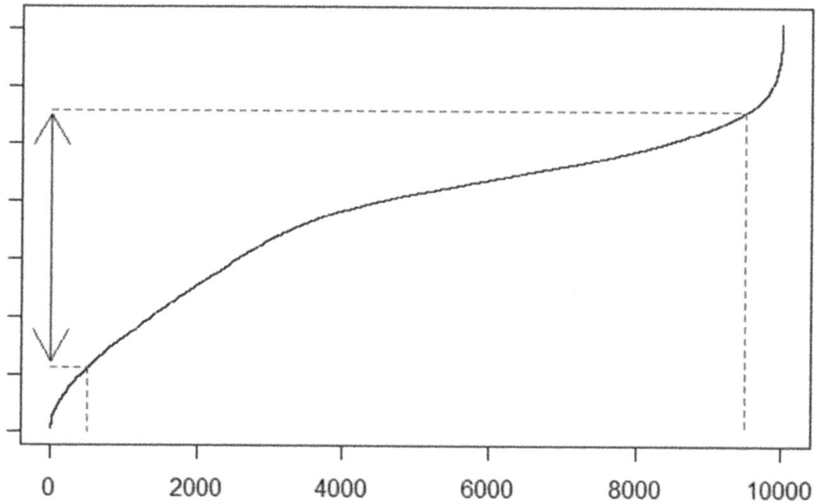

Figure 6.10. *Probability function confidence level. For a color version of this figure, see www.iste.co.uk/bayle/montecarlo.zip*

6.6. Rationale for the choice of the noncentral beta distribution

The following distribution laws, which are likely to have a form of density compatible with the measurements, have also been evaluated. Table 6.1 presents the different distributions and their characteristics (probability density and support).

For the choice of the distribution that most represented the data, we used the Akaike information criterion (AIC) (Akaike 1974). When a statistical model is estimated, it is possible to increase the model likelihood by adding a parameter. This criterion is used to penalize models according to the number of parameters to satisfy parsimony criterion. We then choose the model with the lowest AIC.

The following curve illustrates the estimated and truncated probability densities between 0 and 1, with the AIC displayed in parentheses.

According to the graph above in Figure 6.10, the noncentral beta distribution best corresponds to the data: bell shape + high-value vertex in the measurements and the lowest AIC.

Distribution	Probability density	Support
Beta	$f(x;p,q,a,b) = \frac{1}{B(p,q)}) \frac{(x-a)^{p-1}(b-x)^{q-1}}{(b-a)^{p+q-1}}$ $B(x,y) = \int_0^1 t^{x-1}(1-t)^{y-1}dt = \frac{\Gamma(x)\Gamma(y)}{\Gamma(x+y)}$	$a \leq x \leq b$
Non-central beta	$f(x;\alpha,\beta,\lambda) = \sum_{j=0}^{\infty} \frac{1}{j!}\left(\frac{\lambda}{2}\right)^j e^{-\frac{\lambda}{2}} \frac{x^{\alpha+j-1}(1-x)^{\beta-1}}{B(\alpha+j,\beta)}$	$0 \leq x \leq 1$
Chi-square	$f(x;\nu,\gamma) = \frac{(x-\gamma)^{\frac{\nu}{2}-1} e^{-\frac{x-\gamma}{2}}}{2^{\frac{\nu}{2}}\Gamma\left(\frac{\nu}{2}\right)}$	$\gamma \leq x \leq +\infty$
Student's t distribution	$f(x;n) = \frac{\left(1+\frac{x^2}{n}\right)^{-\frac{n+1}{2}}}{\sqrt{n}.B\left(\frac{1}{2},\frac{n}{2}\right)}$	$-\infty \leq x \leq +\infty$
LogNormal	$f(x;\mu,\sigma,\gamma) = \frac{1}{(x-\gamma)\sigma\sqrt{2\pi}} e^{-\frac{1}{2}\left(\frac{\ln(x-\gamma)-\mu}{\sigma}\right)^2}$	$\gamma \leq x \leq +\infty$
Normal	$f(x;\mu,\sigma) = \frac{1}{\sigma\sqrt{2\pi}} e^{-\frac{1}{2}\left(\frac{x-\mu}{\sigma}\right)^2}$	$-\infty \leq x \leq +\infty$
Weibull	$f(x;\eta,\sigma,\gamma) = \frac{\eta}{\sigma}\left(\frac{x-\gamma}{\sigma}\right)^{\eta-1} e^{-\left(\frac{x-\gamma}{\sigma}\right)^{\eta}}$	$\gamma \leq x \leq +\infty$
Gamma	$f(x;\kappa,\theta,\gamma) = \frac{1}{\Gamma(\kappa)\theta^{\kappa}}(x-\gamma)^{\kappa-1} e^{-\frac{x-\gamma}{\theta}}$	$\gamma \leq x \leq +\infty$
F-distribution	$f(x;m,n) = \frac{1}{xB(m,n)}\sqrt{\frac{(mx)^m n^n}{(mx+n)^{m+n}}}$	$0 \leq x \leq +\infty$
Maxwell	$f(x;\alpha) = \frac{1}{\alpha^3}\sqrt{\frac{2}{\pi}} x^2 e^{-\frac{x^2}{2\alpha^2}}$	$0 \leq x \leq +\infty$
Moyal	$f(x) = \frac{1}{\sqrt{2\pi}} e^{-\frac{1}{2}(x+e^{-x})}$	$-\infty \leq x \leq +\infty$
Rayleigh	$f(x;\alpha,\gamma) = \frac{x-\gamma}{\alpha^2} e^{-\frac{x-\gamma^2}{2\alpha^2}}$	$\gamma \leq x \leq +\infty$
Johnson SB	$f(x;\gamma,\delta,\lambda,\zeta) = \frac{\delta}{\lambda\sqrt{2\pi}z(1-z)} e^{-\frac{1}{2}\left(\gamma+\delta\ln\left(\frac{z}{1-z}\right)\right)^2}$ $z \equiv \frac{x-\zeta}{\lambda}$	$\zeta \leq x \leq \zeta+\lambda$

Table 6.1. *Main probability distributions potentially compatible*

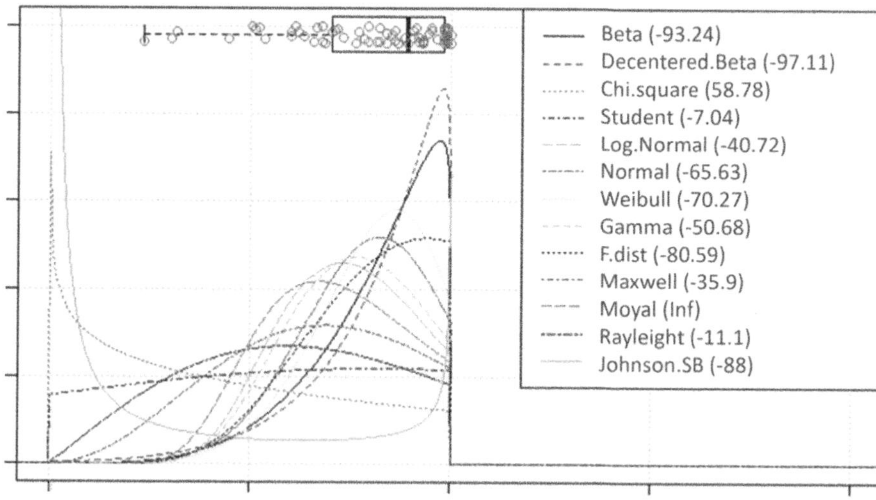

Figure 6.11. *Probability density and AIC for the different distributions selected. For a color version of this figure, see www.iste.co.uk/bayle/montecarlo.zip*

6.7. Summary

– In some specific cases, the usual probability distributions do not always yield satisfactory results. In this chapter, we have presented a real case where the use of the noncentral beta distribution proved necessary.

– The peculiarity of its domain of definition, the interval [0; 1], has been avoided by normalizing the data originating from the measurements.

– The complexity of the expression of its probability of failure led us to use a simulation method other than the inversion method.

– With the maximum likelihood method, we were able to estimate the parameters of the noncentral beta distribution.

– We introduced in detail the rejection method algorithm that allowed us to obtain a confidence interval for the probability of failure of a critical event.

– Finally, using the AIC, we were able to justify the use of this distribution compared with other probability distributions that could be potentially utilized.

7

Statistical Characteristics of HPP and PLP Processes

When a system is subjected to maintenance actions, the concept of one–off processes should be introduced. These processes describe the occurrence of failures over time. When, owing to maintenance actions, the system returns to the same state of reliability as it was before the failure, it is called minimal maintenance, and the failure intensity λ is then a deterministic function of time. This type of maintenance is modeled by Poisson processes (Gaudoin 2007).

7.1. Reminders about Poisson processes

Poisson process theory applies particularly well to complex systems. A point process counting system failure is a Poisson process (Gaudoin 2007) if:

N(0) = 0

For any a < b ≤ c, N[a,b] and N[c,d] are independent.

If a < b, then $N[a,b]$ follows a Poisson distribution of parameter $\int_a^b \lambda(u).du$, where N[a,b] is the number of failures between times "a" and "b".

7.2. HPP homogeneous Poisson process

The homogeneous Poisson process (HPP) (Gaudoin 2007) is a special case of a Poisson process for which the intensity of failure is independent of time. This model is as simple as possible for systems with maintenance. It is well adapted to mature systems, namely, systems in which only catalectic failures are observed.

The failure intensity is therefore written as:

$$\lambda(t) = \lambda$$

The estimation of the parameter λ is achieved based on the maximum likelihood method. Let us consider the case where the process is observed over a fixed time interval. The observation is the number "n" of failures detected at times $t_1, t_2, \ldots, t_n$.

In this case, the log-likelihood function is given by:

$$\mathcal{L}(\lambda, n, t_1, \ldots, t_n) = \text{n}.\ln(\lambda) - \lambda.t$$

By maximizing the likelihood, we obtain an estimator of the parameter λ, that is:

$$\hat{\lambda} = \frac{\text{n}}{\text{t}_n} \qquad [7.1]$$

7.2.1. *Parameter HPP bias*

An estimator of the parameter λ is unbiased if and only if $E[\hat{\lambda}] = \lambda$. On the other hand, we know that the sum of "n" exponential distributions of parameter λ is a gamma distribution of parameters n and λ (Appendix 4). Hence:

$$E[\hat{\lambda}] = \text{E}\left[\frac{\text{n}}{\sum_{i=1}^{n} \text{T}_i}\right] = \int_0^{+\infty} \frac{n}{t}.\frac{\lambda^n}{(n-1)!}.exp(-\lambda.t).t^{n-1}.dt$$

namely:

$$E[\hat{\lambda}] = \frac{n.\lambda^n}{(n-1)!}.\int_0^{+\infty} exp(-\lambda.t).t^{n-2}.dt$$

that is:

$$E[\hat{\lambda}] = \frac{n.\lambda}{(n-1)}.\int_0^{+\infty} \frac{\lambda^{n-1}}{(n-2)!}.exp(-\lambda.t).t^{n-2}.dt$$

The term $\frac{\lambda^{n-1}}{(n-2)!}.exp(-\lambda.t).t^{n-2}$ represents the probability density of a gamma distribution of parameters n-1 and λ.

Consequently, the above integral represents the probability of failure of the gamma distribution and is therefore 1, from which:

$$E[\hat{\lambda}] = \frac{n.\lambda}{n-1}$$

Therefore, the estimator is biased since its expectation is not equal to the theoretical value. An unbiased estimator of the parameter λ would then be:

$$\widehat{\lambda'} = \frac{\mathrm{n}-1}{\mathrm{n}}.\hat{\lambda} = \frac{\mathrm{n}-1}{\sum_{i=1}^{n} \mathrm{T}_i} \qquad [7.2]$$

Therefore, $\hat{\lambda}$ is biased, but the estimator $\widehat{\lambda'} = \left(\frac{n-1}{n}\right).\hat{\lambda}$ is unbiased.

7.2.2. *Coefficient of variation of the HPP parameter*

The variance of $\widehat{\lambda'}$ is given by (Gaudoin 2007):

$$var(\widehat{\lambda'}) = \frac{\lambda^2}{\mathrm{n}-2}$$

Its coefficient of variation is therefore given by:

$$CV(\widehat{\lambda'}) = \frac{E[\widehat{\lambda'}]}{\sqrt{var(\widehat{\lambda'})}} = \frac{\frac{n}{n-1}.\lambda}{\sqrt{\frac{\lambda^2}{\mathrm{n}-2}}} = \frac{n.\sqrt{n-2}}{n-1} \qquad [7.3]$$

7.3. PLP power process

In practice, an increase can be observed in the intensity of failure or, more conventionally, a decrease. The increase is because, despite the maintenance actions carried out during failure, the system deteriorates over time because of different aging mechanisms. A decrease is often observed in the early years of operational activity because design changes occur in the system. However, the HPP model is not suitable. Crow (1974) then proposed a failure intensity model potentially responding to the two cases mentioned above from the power process defined by:

$$\lambda(t) = \alpha.\beta.t^{\beta-1} \quad [7.4]$$

REMARK.–

The parameter of the PLP power process should not be mistaken β for the parameter β of the Weibull distribution. The PLP is part of the family of nonhomogeneous Poisson processes (NHPPs).

This model generalizes the previous model since when $\beta = 1$, the HPP process occurs again. If $\beta < 1$, we indeed have a failure intensity that decreases and that increases when $\beta > 1$, thus modeling all of the case scenarios considered. The parameters of the PLP model are estimated via the maximum likelihood method.

Consider the case where the process is observed over a fixed time interval (0,t). The observation is the number "n" of failures detected at times t_1, t_2..., t_n. In this case, the likelihood-log function is given by (Gaudoin 2007):

$$\mathcal{L}(\alpha, \beta, t_1, \ldots, t_n) = \alpha^n.\beta^n.\prod_{i=1}^{n} t_i{}^{\beta-1}.exp(\alpha.t^{\beta})$$

By deriving this equation with respect to α and β, a system of two equations is obtained from which the corresponding estimators are derived, namely:

$$\hat{\beta} = \frac{\mathrm{n}}{\sum_{i=1}^{n} ln\left(\frac{t_n}{t_i}\right)} \quad [7.5]$$

$$\hat{\alpha} = \frac{n}{t^{\hat{\beta}}} \quad [7.6]$$

Unlike the HPP model, there are no explicit expressions of the bias and variance of the estimators of parameters α and β. As in Chapter 2, we focus on the relative bias defined by equation [2.5] and the coefficient of variation of the two parameters α and β.

These quantities are difficult to compute, and we then propose performing Monte Carlo simulations to estimate them. For this purpose, the number of failures (10, 20, 40, 70 and 100) is varied for different values of the parameter β (0.7; 1; 1.5 and 2).

7.3.1. *Power process simulations*

The simulation of "n" failure instants following a power process is given by the algorithm shown in Figure 7.1.

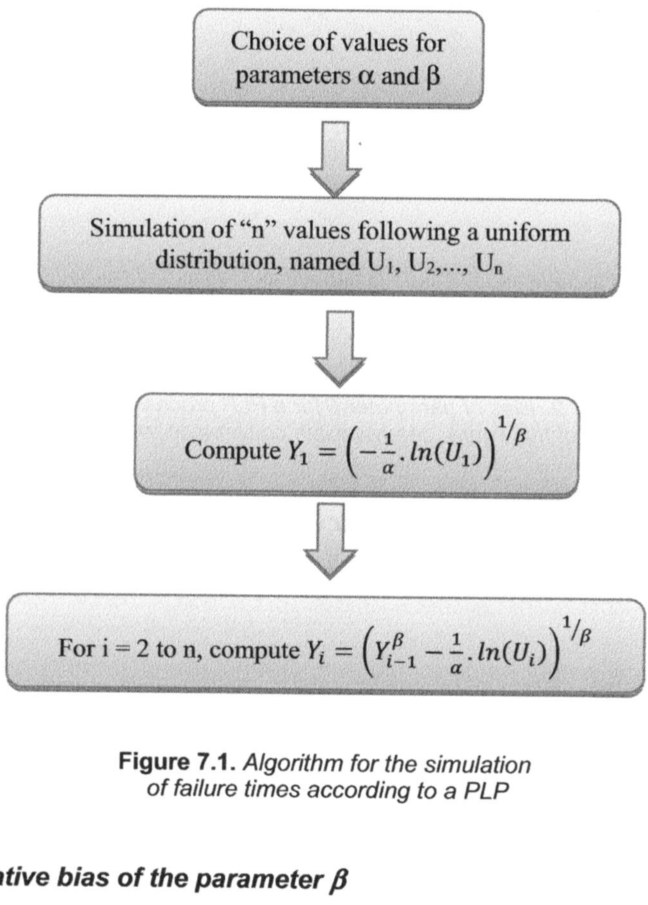

Figure 7.1. *Algorithm for the simulation of failure times according to a PLP*

7.3.2. *Relative bias of the parameter β*

Figure 7.2 shows the relative bias of the parameter β according to the number of failures and its value.

The bias of the parameter β largely depends on the number of failures but does not depend much on its intrinsic value since the curves are practically superimposed.

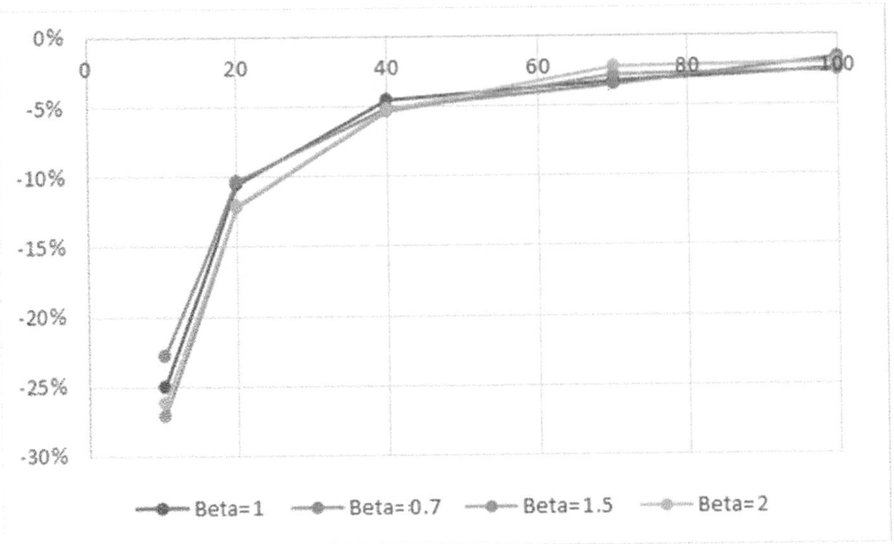

Figure 7.2. *Bias of parameter β for a PLP process. For a color version of this figure, see www.iste.co.uk/bayle/montecarlo.zip*

7.3.3. *Relative bias of the parameter α*

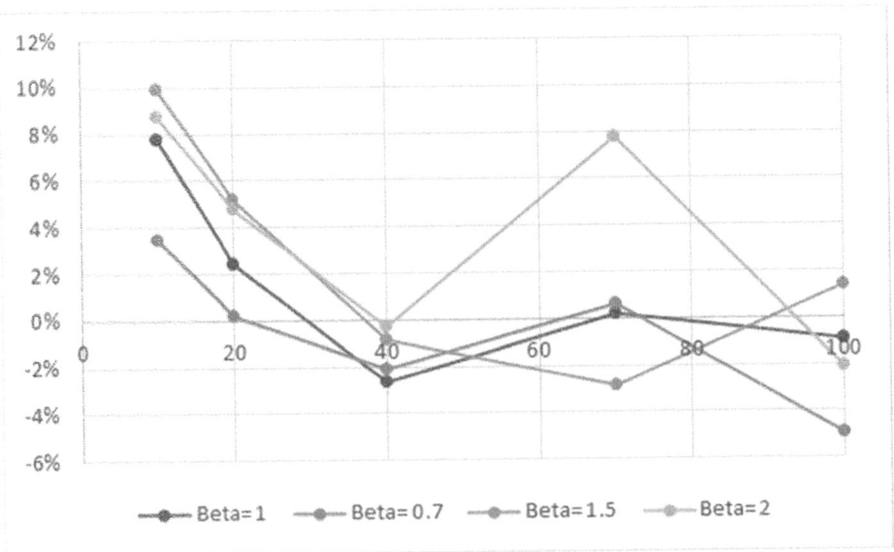

Figure 7.3. *Bias of parameter α for a PLP process. For a color version of this figure, see www.iste.co.uk/bayle/montecarlo.zip*

The bias of the parameter α is quite small and depends little on the number of failures and the value of the parameter β.

7.3.4. *Coefficient of variation of the parameter β*

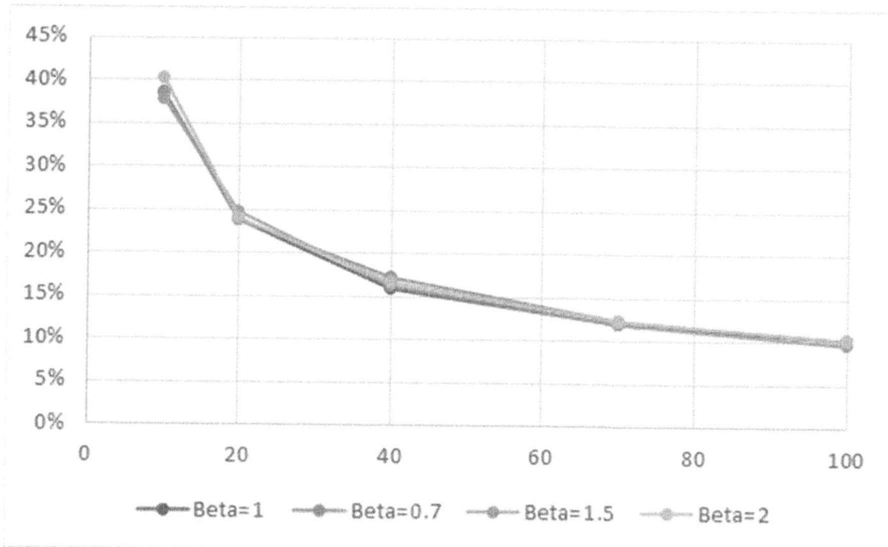

Figure 7.4. *Coefficient of variation of the parameter β for a PLP process. For a color version of this figure, see www.iste.co.uk/bayle/montecarlo.zip*

The coefficient of variation of the parameter β depends on the number of failures and is quite large when the number of failures is low (20).

7.3.5. *Coefficient of variation of the parameter α*

The coefficient of variation of the parameter α depends on the number of failures and is very significant even if the number of failures is substantial. It does not depend much on the parameter β.

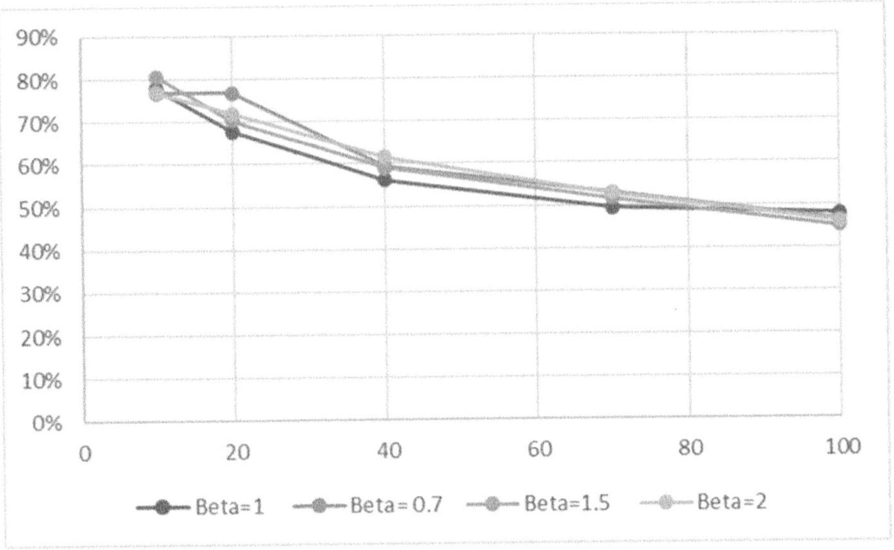

Figure 7.5. *Coefficient of variation of the parameter α for a PLP process. For a color version of this figure, see www.iste.co.uk/bayle/montecarlo.zip*

7.4. Summary

– Bias and variance are the two statistical properties of parameter estimators in a reliability model.

– To better compare the results obtained, we used the relative bias and the coefficient of variation.

– For a homogeneous Poisson process HPP, there are explicit solutions for calculating them.

– For a PLP power process, there are no explicit solutions for the estimation of the parameters η and β. Therefore, neither the relative bias nor the coefficient of variation.

– The relative bias of the parameter β can be significant if the number of failures is small. It does not depend on the values of the parameters β. The maximum likelihood method thus tends to overestimate the value of β.

– The relative bias of the scaling parameter α is negligible, regardless of the number of failures and the value of the parameter β.

– The coefficient of variation of parameter β can be significant if the number of failures is small. It does not depend on the values of parameters α and β.

– The coefficient of variation of the parameter α can be significant because it depends on the number of failures but depends little on the value of β.

Part 2

Maintainability

8

Maintainability

The maintainability of an operational system applies, of course, to systems with maintenance. We will assume that repair times are negligible compared to times of correct operation, such as the theory of renewal processes developed in Chapter 4 is fully applicable. The maintainability of a system consists of its ability to replace an operational system that is failing and therefore to optimally design a stock of systems.

8.1. Average number of failures

In terms of maintenance, the number of failures that will be observed is, of major interest. However, because of the random nature of failure times, it is not possible to precisely know the number of failures at a given time "t" but only an average value.

This is given from equation [3.30], that is:

$$N(t) = \int_0^t Rocof(u).du \qquad [8.1]$$

For more details on the effect of maintenance on reliability and modeling, we advise the reader to refer to Basu and Rigdon (2000) or Bayle (2019). Equation [8.1] indicates an integral relationship between the average number of failures and the Rocof.

However, the latter has been extensively explained in detail and illustrated in Chapter 3 on reliability. As a result, the reader should systematically return to the examples proposed in Chapter 3.

We can also note that, quite quickly, the average number of failures follows a line of slope 1/MTTF. This is because the Rocof has an asymptotic value of 1/MTTF. In addition, the Rocof oscillates around the asymptotic value depending on the value of the shape parameter of the distribution being considered (parameter β for a Weibull distribution, for example). The properties of the integration cause this oscillation, such that the average number of failures is much lower.

8.2. Serial system

8.2.1. *Element-level renewal*

We observed in Chapter 4 that, for a serial system, we have an overlapping renewal process.

From equations [3.35] and [8.1], we can write that:

$$N(t) = \int_0^t \sum_{i=1}^{s} Rocof_i(u)\,.\,du$$

namely:

$$N(t) = \sum_{i=1}^{s} \int_0^t Rocof_i(u).\,du \qquad [8.2]$$

Exponential distribution

The Rocof for an exponential parameter distribution λ is equal to λ.

Equation [8.2] is therefore written as: $N(t) = \sum_{i=1}^{s} \int_0^t \lambda_i.\,du$ or furthermore:

$$N(t) = \sum_{i=1}^{s} \lambda_i.\,t \qquad [8.3]$$

Normal distribution

The Rocof of a serial system whose elements are normally distributed:

$$Rocof(t) = \sum_{i=1}^{+\infty} \phi(u, \mu.\,i, \sigma.\sqrt{i})$$

The average number of failures at time "t" is obtained by integrating equation [3.39] as follows:

$$N(t) = \int_{0}^{t} \sum_{i=1}^{S} \sum_{i=1}^{+\infty} \phi\left(u, \mu . i, \sigma . \sqrt{i}\right) . du$$

and finally:

$$N(t) = \sum_{i=1}^{S} \sum_{i=1}^{+\infty} \phi\left(u, \mu . i, \sigma . \sqrt{i}\right) \qquad [8.4]$$

If we reuse the inputs from example 3.4: Element 1: Normal(500; 100) Element 1: Normal(1,000; 150), we get the results in Figure 8.1.

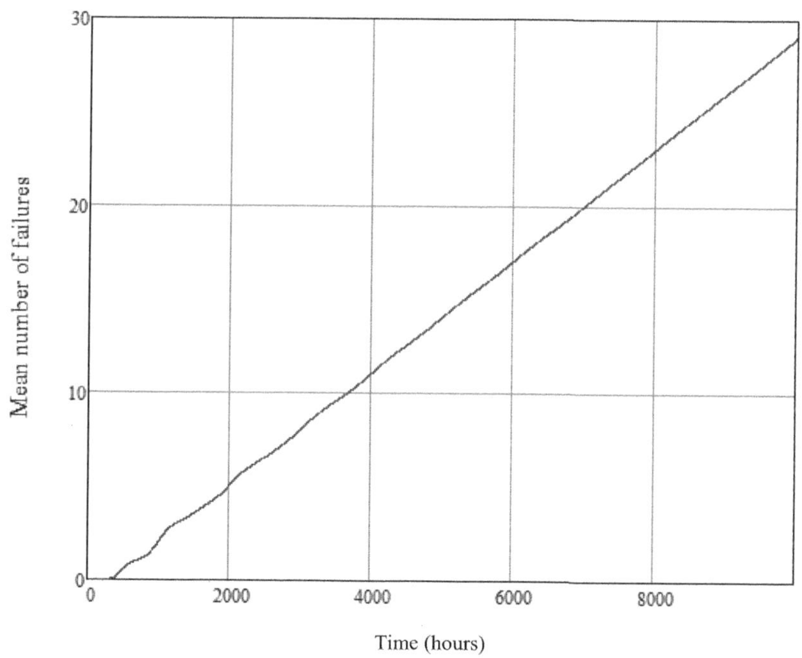

Figure 8.1. *Mean failures number for two normal distributions in series with maintenance at element level. For a color version of this figure, see www.iste.co.uk/bayle/montecarlo.zip*

Simulations

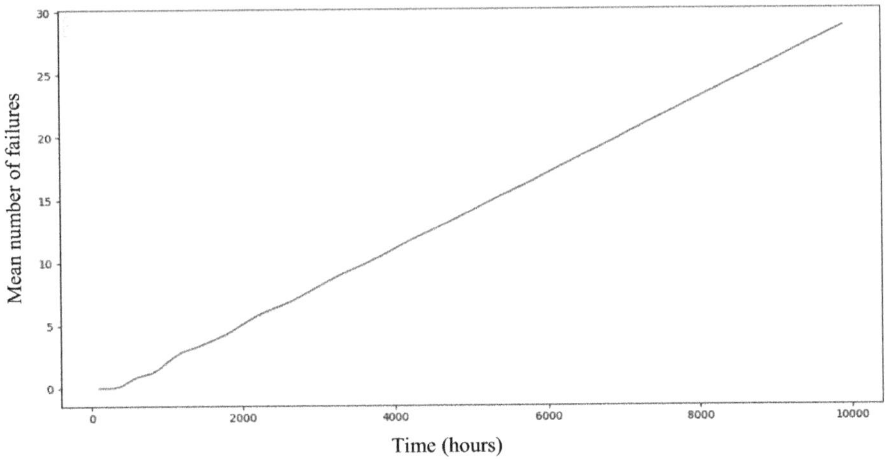

Figure 8.2. *Simulation of th mean failures number for two normal distributions in series with maintenance at element level. For a color version of this figure, see www.iste.co.uk/bayle/montecarlo.zip*

Weibull distribution

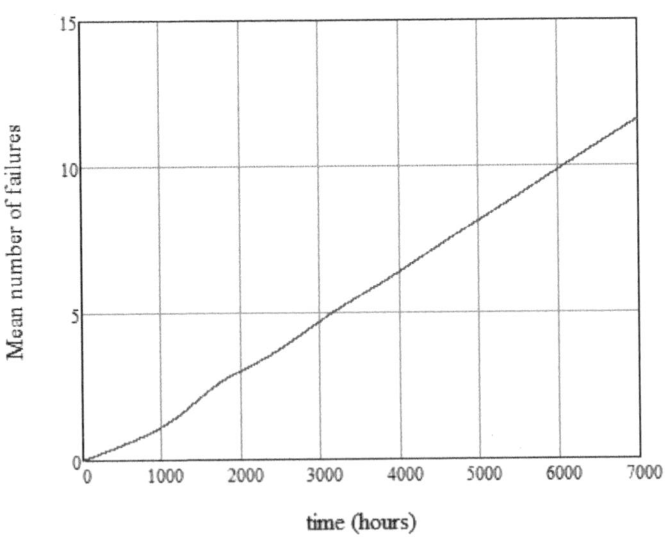

Figure 8.3. *Mean failures number for two Weibull distributions in series with maintenance at element level. For a color version of this figure, see www.iste.co.uk/bayle/montecarlo.zip*

We know that the Rocof for a Weibull distribution has no explicit solutions. Consequently, the Rocof in a serial system with elements according to a Weibull distribution does not have any either. Reusing the inputs from example 3.5, we obtain the following via computation.

Simulation

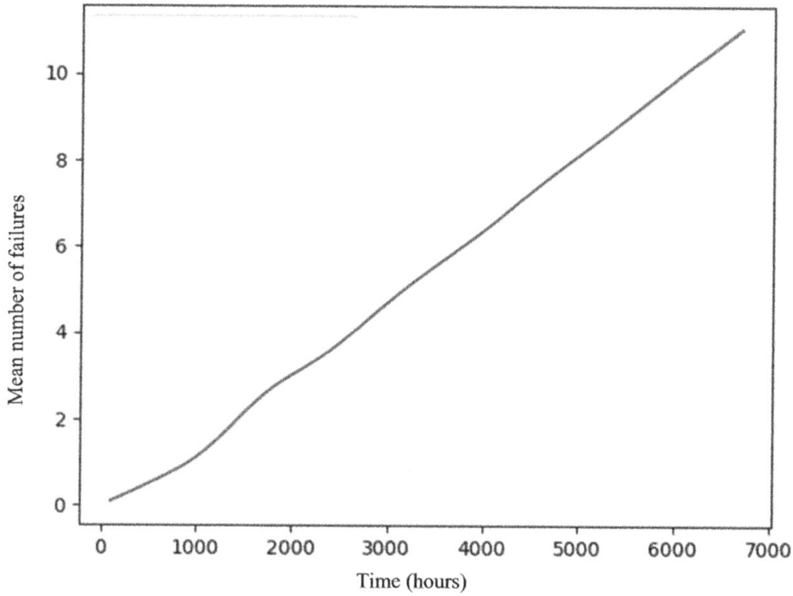

Figure 8.4. *Simulation of the mean failures number for two Weibull distributions in series with maintenance at element level. For a color version of this figure, see www.iste.co.uk/bayle/montecarlo.zip*

8.2.2. *System-level renewal*

We observed in section 3.3.2.2 that the Rocof for a serial system with system-level renewal has no explicit solutions except for exponential distributions. Consequently, only numerical computation is possible.

Using the inputs from example 3.6 (consider a problem with a system comprising two elements with the following characteristics: Element 1: Normal(500; 100) Element 2: Normal(1,000; 150), the average number of failures is illustrated by Figure 8.5.

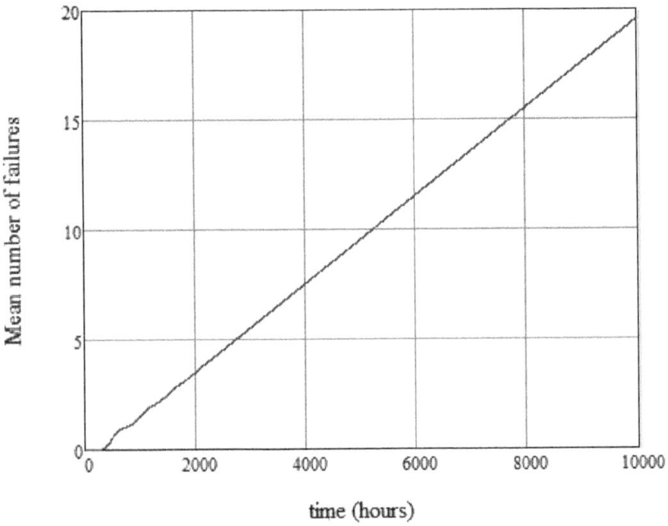

Figure 8.5. *Mean failures number for two normal distributions in series with maintenance at system level. For a color version of this figure, see www.iste.co.uk/bayle/montecarlo.zip*

Simulation

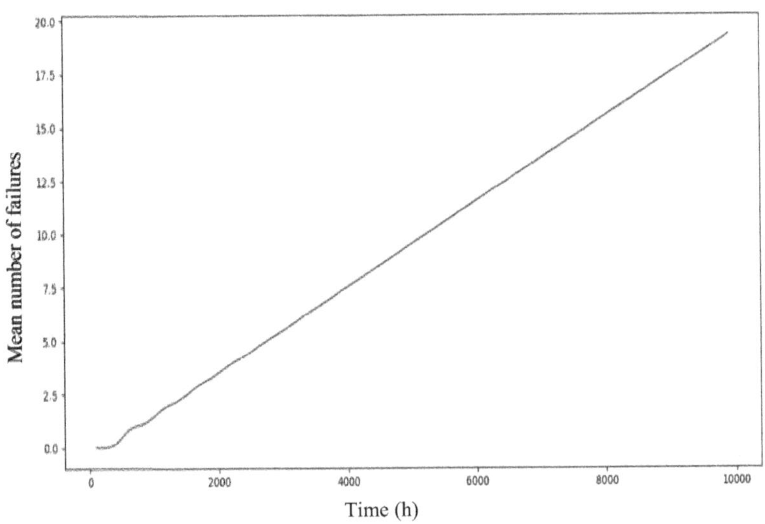

Figure 8.6. *Simulation of the mean failures number for two normal distributions in series with maintenance at system level. For a color version of this figure, see www.iste.co.uk/bayle/montecarlo.zip*

This approach may help us to compare the two types of (element- or system-based) maintenance for the problem with the two normal distributions of example 3.4. This is shown in Figure 8.7.

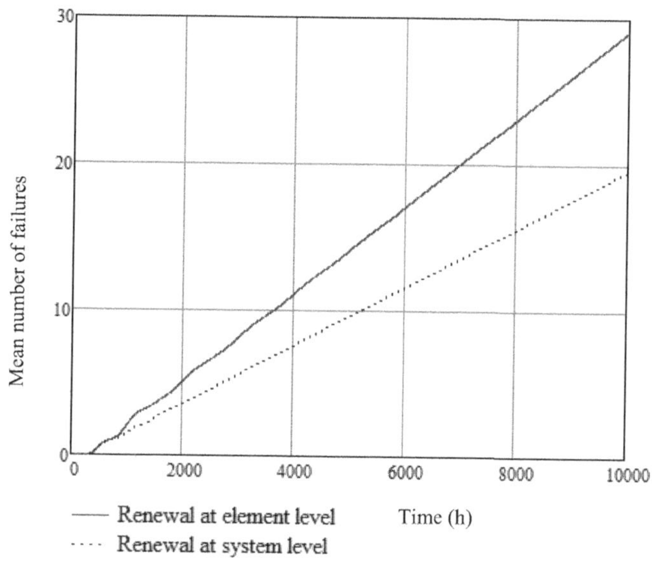

Figure 8.7. *Comparison of the mean number of failures at element and system levels. For a color version of this figure, see www.iste.co.uk/bayle/montecarlo.zip*

It can be clearly seen that the number of failures with system-level renewal (blue curve) is lower. Therefore, this strategy improves the system reliability level at the expense of costs.

8.3. Parallel system

For element-level renewal, we observed in section 3.3.3 that the Rocof was defined by equation [3.41], that is:

$$Rocof_P(t) = \sum_{j=1}^{+\infty} \left(p.\mathrm{F(t)}^{\mathrm{p}-1}.\mathrm{f(t)}\right)^{<j>}$$

Therefore, the average number of failures is given by:

$$N(t) = \int_0^t \sum_{j=1}^{+\infty} \left(\mathrm{p.F(u)^{p-1}.f(u)}\right)^{<j>} \qquad [8.5]$$

Exponential distribution

Since the previous equation gives the average number of failures for exponential distributions, it has no explicit solutions.

If we revisit example 3.7 for three Weibull distributions of parameters λ = 0.001, we get the results in Figure 8.8 for the numerical computation.

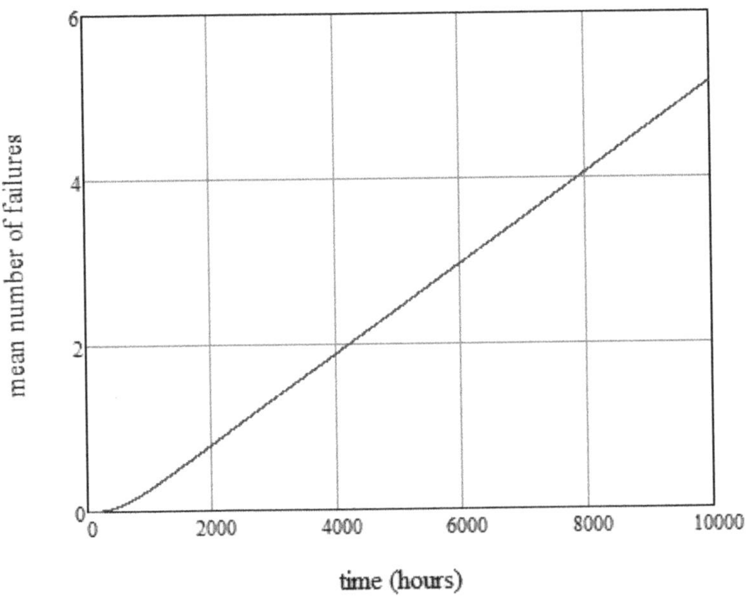

Figure 8.8. *Mean failures number for exponential distributions in parallel with maintenance at element level. For a color version of this figure, see www.iste.co.uk/bayle/montecarlo.zip*

Simulation

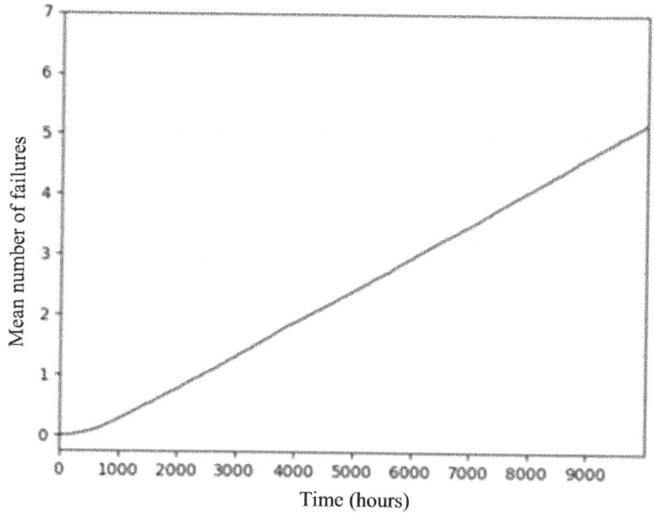

Figure 8.9. *Simulation of the mean failures number for exponential distributions in parallel with maintenance at element level. For a color version of this figure, see www.iste.co.uk/bayle/montecarlo.zip*

Weibull distributions

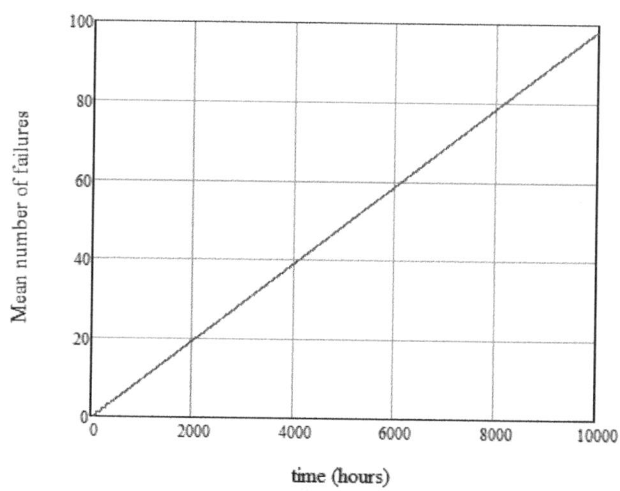

Figure 8.10. *Mean failures number for Weibull distributions in parallel with maintenance at element level. For a color version of this figure, see www.iste.co.uk/bayle/montecarlo.zip*

Similarly, the Rocof for a parallel system with element-level renewal has no explicit solutions. If we reconsider for two Weibull distributions of parameters $\eta = 100$ and $\beta = 7$, we get the results of Figure 8.10 for the numerical computation.

Simulation

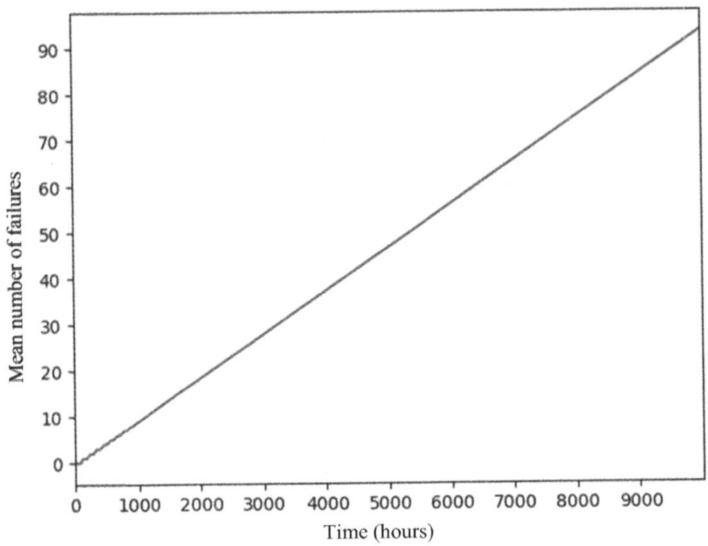

Figure 8.11. *Simulation of the mean failures number for Weibull distributions in parallel with maintenance at element level. For a color version of this figure, see www.iste.co.uk/bayle/montecarlo.zip*

We recall here that, for a parallel system, at the element or system level, maintenance is identical since the system is faulty if and only if all of the elements are faulty.

8.4. k/n system

For element-level renewal, in section 3.3.4.1, the Rocof did not have explicit solutions and as a result, neither did the average number of failures.

Exponential distributions

Even for exponential distributions, the average number of failures has no explicit solutions, and numerical computation is needed. If we reuse the inputs from ($\lambda = 0.0001$ n = 4 and k = 2), we get the results in Figure 8.12.

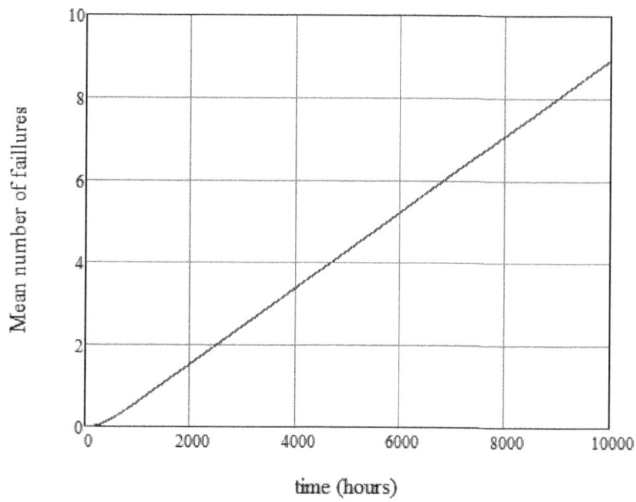

Figure 8.12. *Mean failures number for exponential distributions in k/n redundancy with maintenance at element level. For a color version of this figure, see www.iste.co.uk/bayle/montecarlo.zip*

Simulation

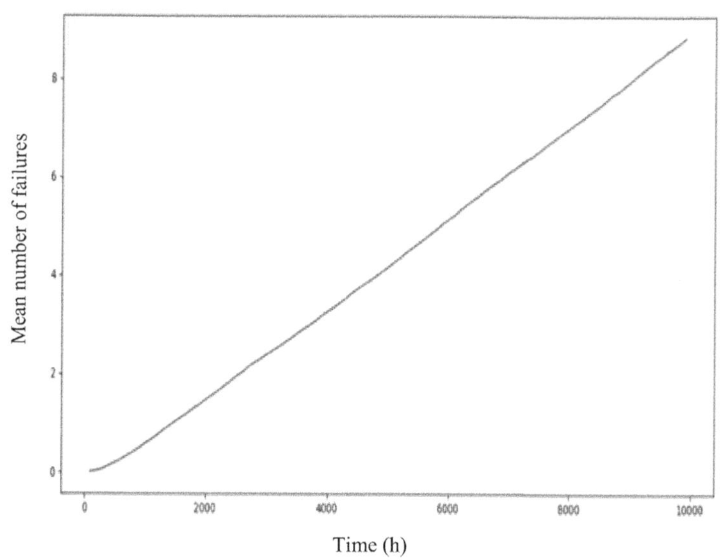

Figure 8.13. *Simulation of the mean failures number for exponential distributions in k/n redundancy with maintenance at element level. For a color version of this figure, see www.iste.co.uk/bayle/montecarlo.zip*

8.5. Avionics system

Element replacement

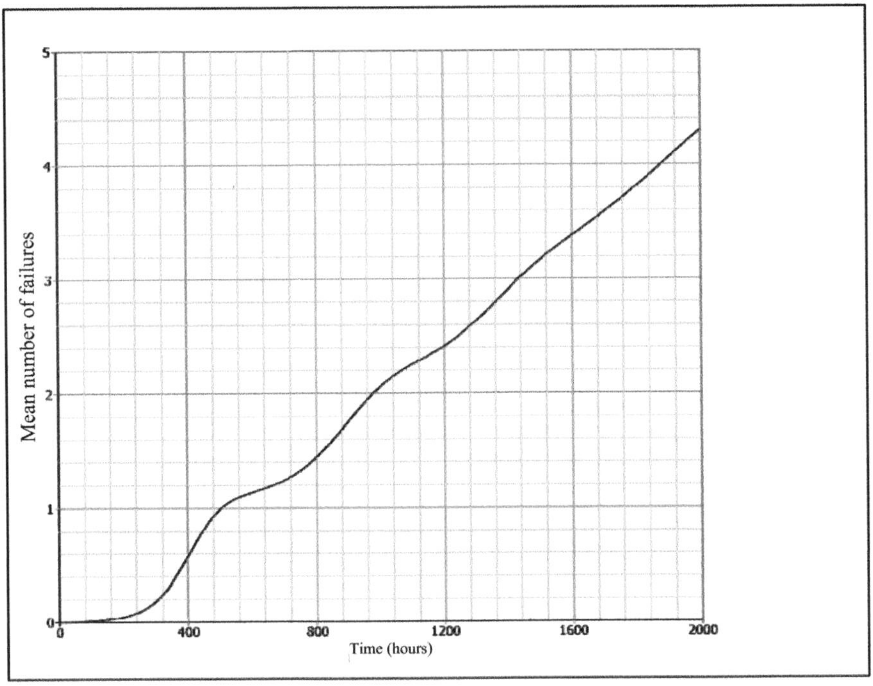

Figure 8.14. *Simulation of the mean failures number in avionic system with maintenance at element level*

System replacement

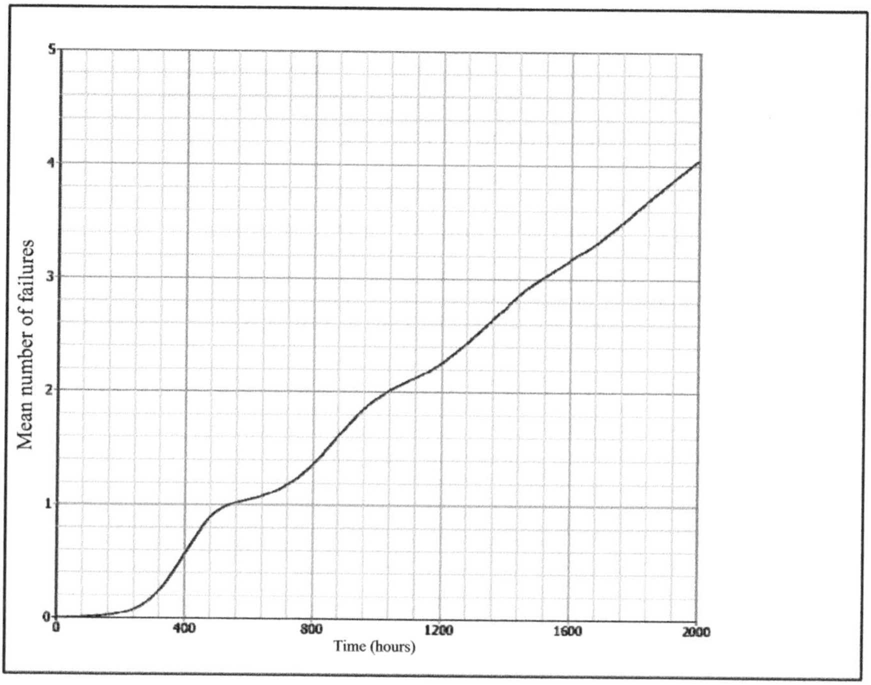

Figure 8.15. *Mean number of failures versus time for the avionic system*

8.6. Summary

The average number of failures N(t) of a system can be expressed on the basis of the Rocof as the failure occurrence rate according to the equation $N(t) = \int_0^t Rocof(u).du$. As such, the results obtained in this chapter are derived from those in Chapter 3 on system reliability.

Serial system:

– For *element-level renewal*, we observed in Chapter 3 that the system Rocof is equal to the sum of the Rocof of each element. Consequently, the average number of system failures is equal to the sum of the average number of failures of each element. If all system elements follow an exponential distribution, then the average number of system failures is written as $N(t) = \sum_{i=1}^{s} \lambda_i.t$. There are nevertheless no explicit solutions as soon as an element follows a Weibull distribution.

– For *system-level renewal*, there are no explicit solutions to the average number of failures unless all of the elements follow an exponential distribution. This number is then the same as the one defined with element-based renewal.

Considering the *other system structures* (parallel, k/n redundancy, series/parallel and parallel/series systems), we should consider the results of Chapter 3 on the Rocof and compute the average number of failures.

We show that $lim_{t \to +\infty} \frac{N(t)}{t} = \frac{1}{\text{MTTF}}$. This result explains why we have a linear evolution of the average number of failures for large time values.

PART 3

Availability

9

System Availability

In Chapter 3, we review the reliability aspects for different types of strategy systems. We made the implicit assumption that repair times were zero, which, from a theoretical point of view, facilitated the use of the properties of the renewal processes. In practice, the operation of a system is a succession of periods of malfunction and nonoperational or repair periods. Additionally, in this chapter, we study the impact of nonzero repair times on another operation quantity and availability.

9.1. Assumptions

We consider the particular context of a system with the following characteristics:

– it is assumed that the system comprises "n" elements and that these are independent of one another;

– we do not consider systems with multiple failures.

When a system fails, a maintenance (repair) action is performed to be functional again. The state of the system with respect to time can be described by a state variable Θ defined by:

$$\Theta(t) = \begin{cases} 1 \text{ if the system is operational at time } t \\ 0 \text{ otherwise} \end{cases}$$

Therefore, the availability A at time t of a system is the probability that it is functional at time t, that is:

$$A(t) = P[\Theta(t) = 1]$$

For a color version of all of the figures in this chapter, see www.iste.co.uk/bayle/montecarlo.zip.

Notably, if the system is not repaired, then the availability is equivalent to the reliability function. Reliability measures a time period, whereas availability is instantaneous (regardless of what happened before t). As long as the system has not yet failed, A(t) = R(t).

Therefore, for non-repairable systems, reliability and availability are equivalent. Unlike R(t), the direction of variation of A(t) is not determined in advance. Most of the time, A(t) admits a limit when t tends to infinity (the asymptotic availability), equal to the average proportion of the time during which the system is in a working state.

Let us assume that uptimes TTF are independent and identically distributed, and that repair times TTR are also independent and identically distributed. Then, TTF + TTR times are independent.

We arrive at the following illustration:

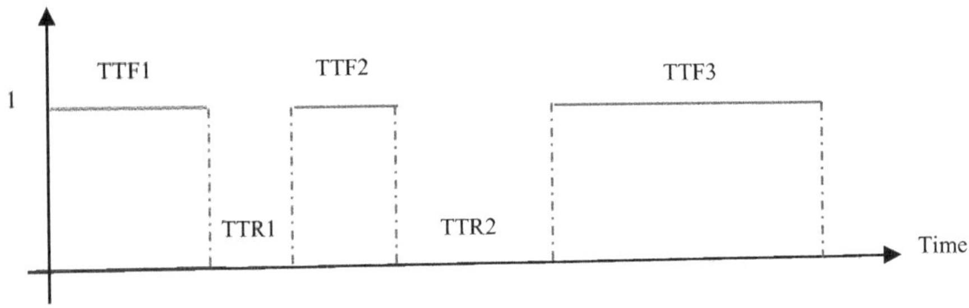

Figure 9.1. *Illustration of times to failure and times to repair*

The average availability is, in general, the main focus. According to the law of large numbers, the arithmetic average tends toward the mathematical expectation such that if n is the observed number of failures, then:

$$\frac{\sum_{i=1}^{n} TTF_i}{n} = E(T) \text{ and } \frac{\sum_{i=1}^{n} TTR_i}{n} = E(D)$$

The time proportion during which the system operates is given by:

$$\frac{\frac{\sum_{i=1}^{n} TTF_i}{n}}{\frac{\sum_{i=1}^{n} TTF_i}{n} + \frac{\sum_{i=1}^{n} TTR_i}{n}} = \frac{E(T)}{E(T) + E(D)} = \frac{MTTF}{MTTF + MTTR} = \frac{MTTF}{MTBF}$$

REMARK.–

Notably, the notion of the MTBF (the true notion of time between successive failures) differs from the notion of the MTBF that is often used in industry to estimate the reliability of a system. In fact, reliability requirements require increasingly larger MTBFs. If we use the true definition, increasing the MTBF would be easy since it would be enough to increase the repair times (MTTR).

In a general framework, it can be shown (Rausand and Hoyland 2004) that the availability of a system is given by:

$$A^*(p) = \frac{1-f_B^*(p)}{\text{p}.\left(1-f_B^*(p).f_R^*(p)\right)} \qquad [9.1]$$

where:

– f_B is the probability density of uptimes;

– f_R is the probability density of repair times;

– p is the operator of the Laplace transform;

– g* Laplace transform of function g.

There are generally no explicit solutions to availability A(t), obtained by inverse transform in the previous equation. This is why Monte Carlo simulations are needed. Nevertheless, there are a few explicit solutions, but in general, they do not translate the operational reality.

9.2. Uptime and repair time: exponential distributions

Consider a system in which uptimes are independent and distributed according to an exponential distribution of parameter λ. Suppose also that repair times are independent and follow an exponential distribution of parameter μ.

Their respective probability densities are therefore valid for $t > 0$:

$$f_B(t) = \lambda.\exp(-\lambda.t)$$

$$f_r(t) = \mu.\exp(-\mu.t)$$

Taking the Laplace transform, we obtain "p" as the Laplace operator:

$$f_B^*(p) = \frac{\lambda}{p+\lambda}$$

$$f_R^*(p) = \frac{\mu}{p + \mu}$$

From equation [9.1], availability can be obtained from the following equation:

$$A^*(p) = \frac{1 - \frac{\lambda}{p + \lambda}}{\text{p}.\left(1 - \left(\frac{\lambda}{p + \lambda}\right).\left(\frac{\mu}{p + \mu}\right)\right)}$$

or furthermore:

$$A^*(p) = \frac{\mu}{p.(\lambda + \mu)} + \left(\frac{\lambda}{\lambda + \mu}\right).\left(\frac{1}{p + (\lambda + \mu)}\right)$$

Taking the inverse Laplace transform, we obtain:

$$A(t) = \frac{\mu}{\lambda + \mu} + \left(A(0) - \frac{\mu}{\lambda + \mu}\right).exp(-(\lambda + \mu).t)$$

In general, the system is functional and therefore available at time t = 0, from which A(0) = 1. Availability can therefore be written as follows:

$$A(t) = \frac{\mu}{\lambda+\mu} + \left(\frac{\lambda}{\lambda+\mu}\right).exp(-(\lambda + \mu).t) \qquad [9.2]$$

Consequently, the asymptotic availability is equal to:

$$A_\infty(t) = lim_{t\to+\infty}\text{A}(t) = \frac{\mu}{\lambda+\mu} = \frac{1/\lambda}{1/\lambda + 1/\mu} = \frac{MTTF}{MTTF+MTTR} \qquad [9.3]$$

REMARK.–

Convergence toward the asymptotic value is very fast in this case since it is exponential. For other distributions, Monte Carlo simulations must be employed to estimate asymptotic availability.

On the other hand, the average number of failures can also be obtained from the following equation:

$$\Lambda^*(p) = \frac{f_B^*(p).f_B^*(p)}{\text{p}.(1 - f_B^*(p).f_R^*(p))}$$

Therefore, for this particular problem, we obtain:

$$\Lambda^*(p) = \frac{\frac{\lambda}{p+\lambda}\cdot\left(\frac{\mu}{p+\mu}\right)}{\text{p.}\left(1-\frac{\lambda}{p+\lambda}\cdot\left(\frac{\mu}{p+\mu}\right)\right)}$$

or still:

$$\Lambda^*(p) = \frac{\lambda.\mu}{p^2.(\lambda+\mu)} - \frac{\lambda.\mu}{p.(\lambda+\mu)^2} + \left(\frac{\lambda.\mu}{(\lambda+\mu)^2}\right)\cdot\left(\frac{1}{p.(\lambda+\mu)}\right)$$

Taking the inverse Laplace transform, we obtain:

$$\Lambda(t) = \left(\frac{\lambda.\mu}{\lambda+\mu}\right).t - \frac{\lambda.\mu}{(\lambda+\mu)^2} + \left(\frac{\lambda.\mu}{(\lambda+\mu)^2}\right).\exp(-(\lambda+\mu).t)$$

EXAMPLE 9.1.– Failure times → exponential distribution $\lambda = 0.001$ and repair times → exponential distribution $\mu = 0.01$. We get the results shown in Figure 9.2.

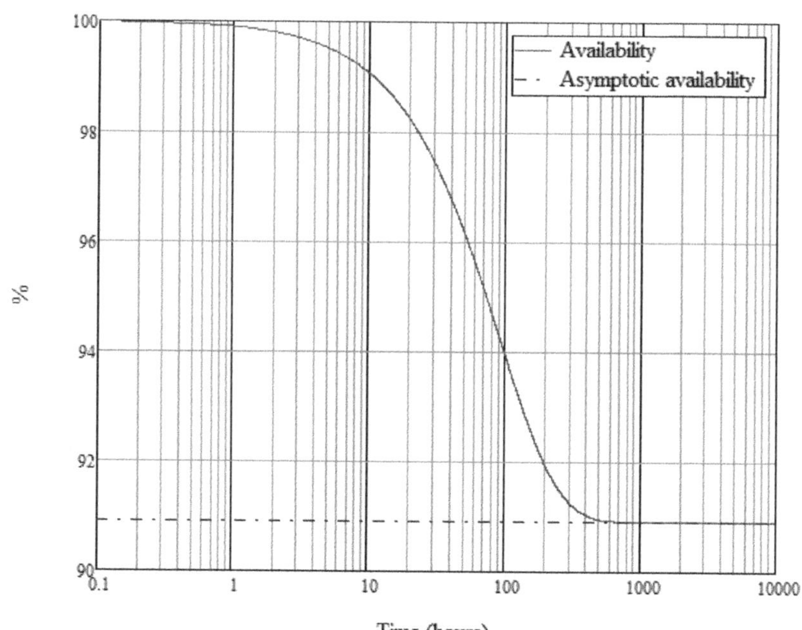

Figure 9.2. *Availability of exponential distributions for failure and repair operations*

9.3. Exponential distribution uptime and constant repair time

Consider a system in which uptimes are independent and distributed according to an exponential distribution of parameter λ. Let us also assume that repair times are constant and equal to ρ. It can then be shown (Rausand and Hoyland 2004) that system availability is given by:

$$A(t) = \sum_{i=0}^{+\infty} \frac{\lambda^i}{i!} . (t - i.\rho)^i . exp(-\lambda.(t - i.\rho)).u(t - i.\rho) \quad [9.4]$$

where u(t) is the rectangle function defined by:

$u(t) = 1$ if $t > 0$ and 0 otherwise

Consequently, we show that the asymptotic availability (Rausand and Hoyland 2004) is:

$$A_\infty = lim_{t \to +\infty} A(t) = \frac{1}{1+\lambda.\rho} \quad [9.5]$$

EXAMPLE 9.2.– Exponential distribution λ = 0.005 and constant repair time ρ = 1,000.

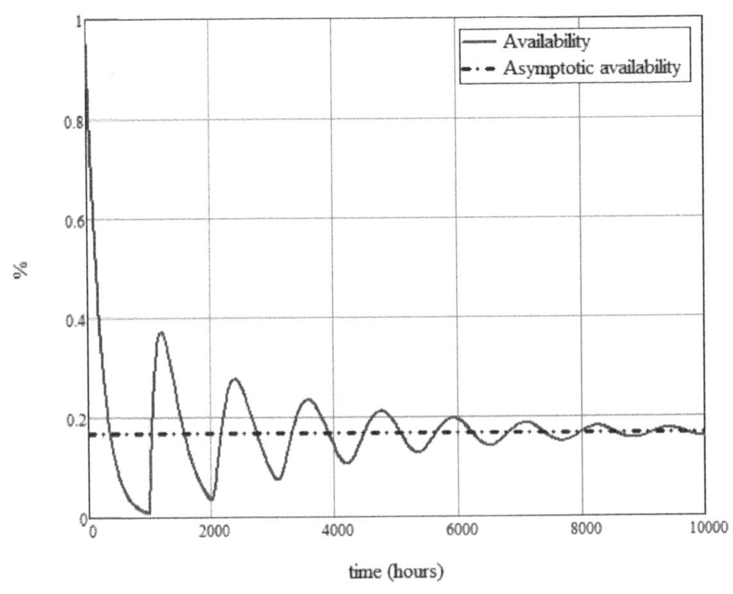

Figure 9.3. *Availability function of example 9.2*

Simulation with BlockSim software

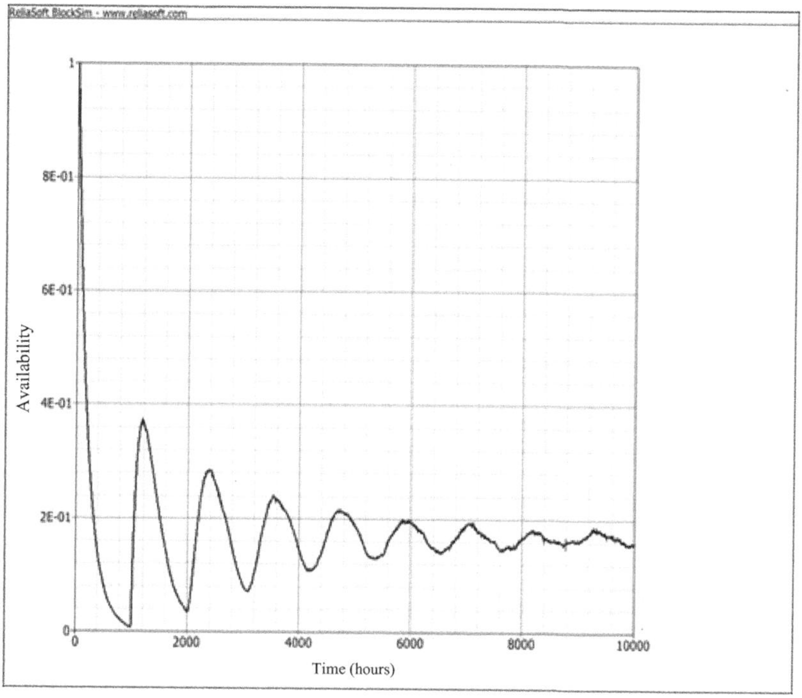

Figure 9.4. *Simulation of availability function of example 9.2*

It takes 10,000 simulations to obtain a plot with acceptable simulation noise. The asymptotic availability is obtained from equation [9.5], that is $A_\infty = \frac{1}{1+\lambda.\rho} = \frac{1}{1+0.004\times 1{,}000} \cong 16\%$.

Availability/unavailability scheme

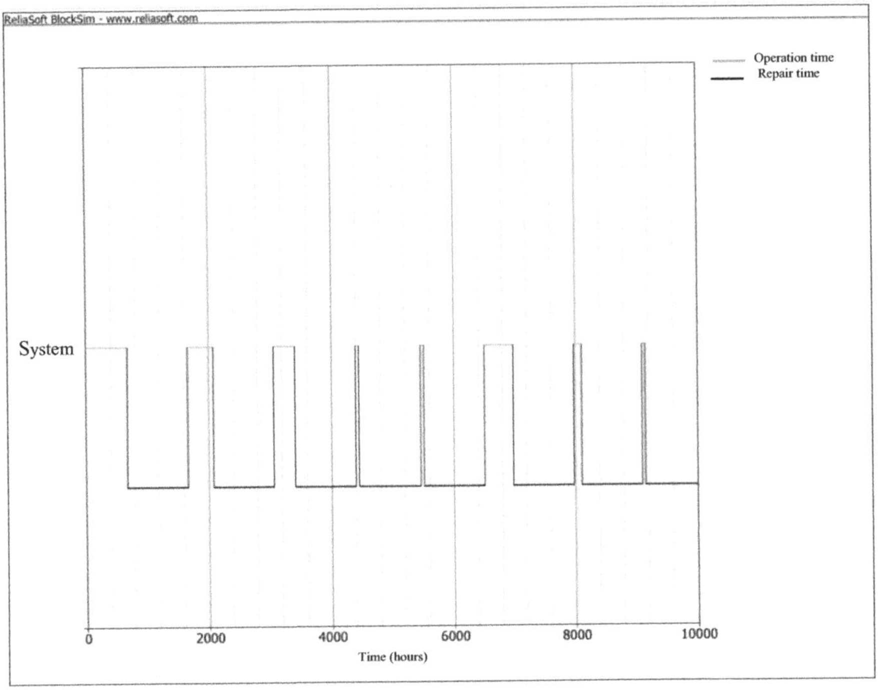

Figure 9.5. *Availability/unavailability scheme of example 9.2*

9.4. Exponential distribution uptime and uniform distribution repair time

In practice, it is unlikely that repair times are constant because it takes more or less time depending on the type of failure observed. There are some cases in which repair times are not known precisely, but only the endpoints of a range of possible values are known. The repair times can then be modeled by a uniform distribution of parameters a and b. The probability density of a uniform distribution is given by:

$$fU(t) = \frac{1}{\mathrm{b} - \mathrm{a}} if t \in [a; b] \text{ and 0 otherwise}$$

Its Laplace transform is then equal to:

$$fU^{*}(p) = \int_{0}^{+\infty} \exp(-p.t).U(t).dt$$

that is:

$$fU^*(p) = \frac{1}{b-a} \cdot \int_a^b \exp(-p.t).dt$$

that is, finally:

$$fU^*(p) = \frac{\exp(-a.p) - \exp(-b.p)}{b-a}$$

Injecting this result into equation [9.1], we obtain:

$$A^*(p) = \frac{1-\frac{\lambda}{p+\lambda}}{p.\left(1-\left(\frac{\lambda}{p+\lambda}\right).\left(\frac{\exp(-a.p)-\exp(-b.p)}{b-a}\right)\right)} \qquad [9.6]$$

To find availability with respect to time, the inverse transform of the previous equation would have to be taken. Unfortunately, there are no known analytical solution to this question, and only Monte Carlo simulations can provide an answer.

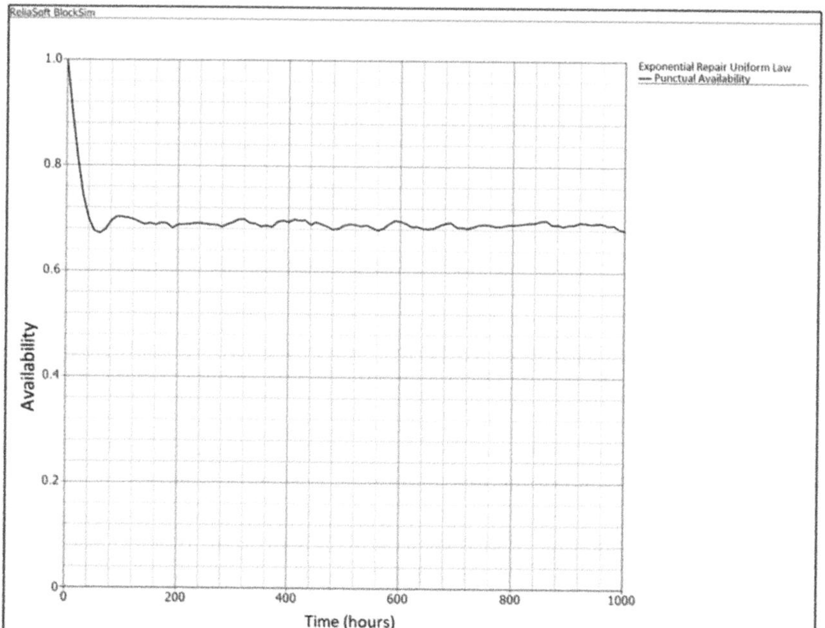

Figure 9.6. *Availability function of example 9.3*

EXAMPLE 9.3.– Exponential distribution $\lambda = 0.005$ and uniform distribution repair time [10, 70].

Asymptotic availability is equal to ~ 0.7.

Availability/unavailability scheme

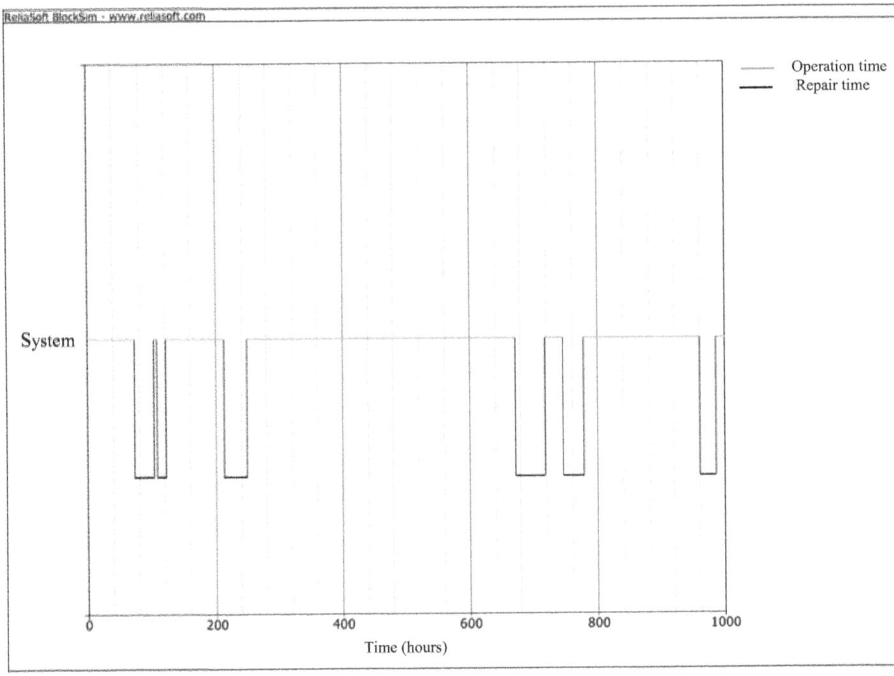

Figure 9.7. *Availability/unavailability scheme of example 9.3*

9.5. Exponential distribution uptimes and normal distribution repair times

It is possible that for some applications, repair times can be systematically recorded and modeled by a normal distribution.

EXAMPLE 9.4.– Exponential distribution $\lambda = 0.005$ and repair times normal distribution with mean $\mu = 50$ and standard deviation $\sigma = 20$. We obtain the results in Figure 9.8.

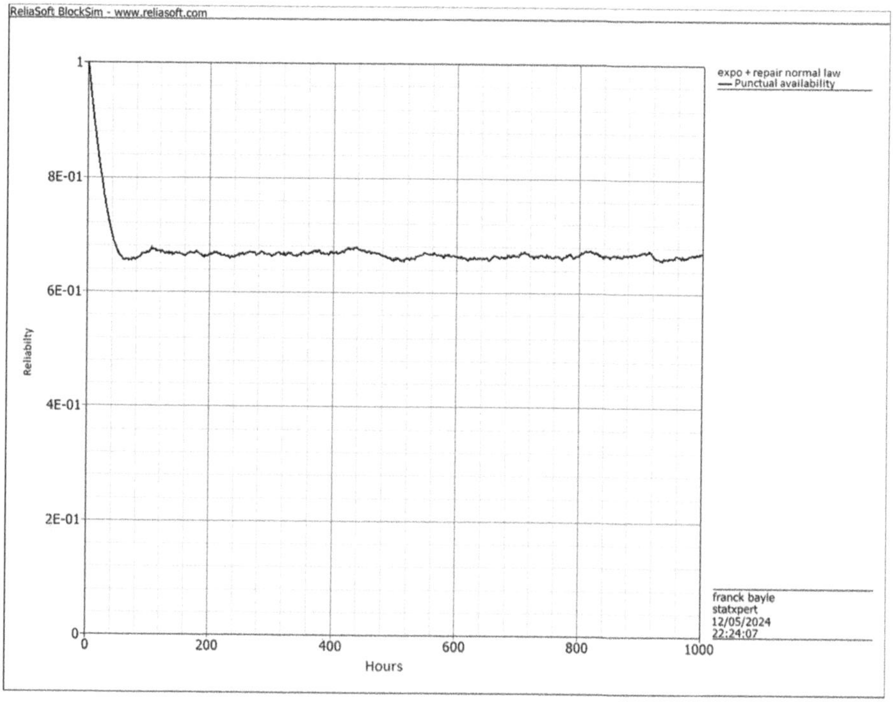

Figure 9.8. *Availability function of example 9.4*

Asymptotic availability is equal to ~ 0.68.

Availability/unavailability scheme

The normal distribution previously used can create two problems:

– It is defined for negative times, which does not make sense for availability.

– It is symmetrical around the mean, and there is no reason for this to be the case in practice. Furthermore, we can substitute a skewed distribution such as the Weibull distribution for particular values of the shape parameter β.

The coefficient of skewness of a Weibull distribution is given by:

$$\xi = \frac{\eta^3.\Gamma\left(1+\frac{3}{\beta}\right) - 3.\mu.\sigma^2 - \mu^3}{\sigma^3}$$

where

$$\mu = \eta . \Gamma\left(1 + \frac{1}{\beta}\right)$$

represents the mean and where

$$\sigma = \sqrt{\eta^2 . \Gamma\left(1 + \frac{2}{\beta}\right)^2 - \left(\eta . \Gamma\left(1 + \frac{1}{\beta}\right)\right)^2}$$

represents the standard deviation of a Weibull distribution.

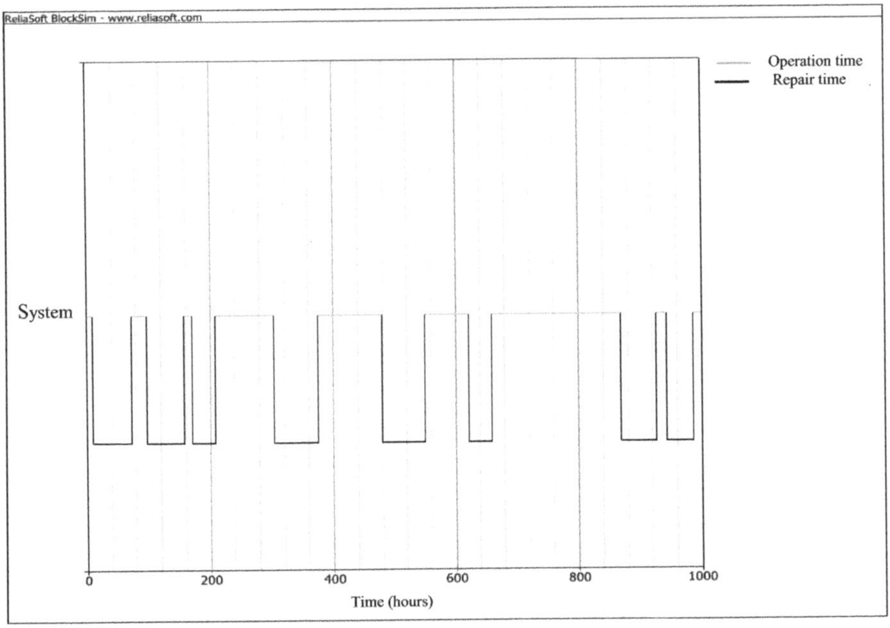

Figure 9.9. *Availability/unavailability scheme of example 9.4*

When the skewness coefficient is negative, the spread of the values is on the left. When the skewness coefficient is positive, the spread of the values is on the right.

The value of the parameter β (independent of the value of the scale parameter η), for which this coefficient is 0, is ~ 3.6.

If the evolution of the skewness coefficient is plotted with respect to the parameter β, we obtain the results in Figure 9.10.

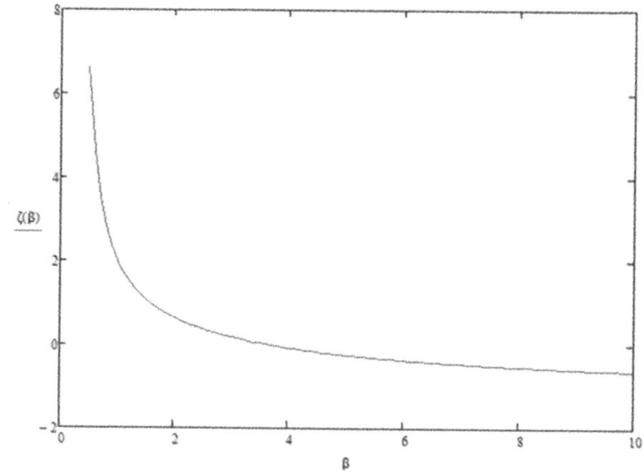

Figure 9.10. *Skewness parameter versus beta parameter*

We see that the probability density of a Weibull distribution is strongly skewed for values of β slightly greater than 1 (e.g. for β = 1.3).

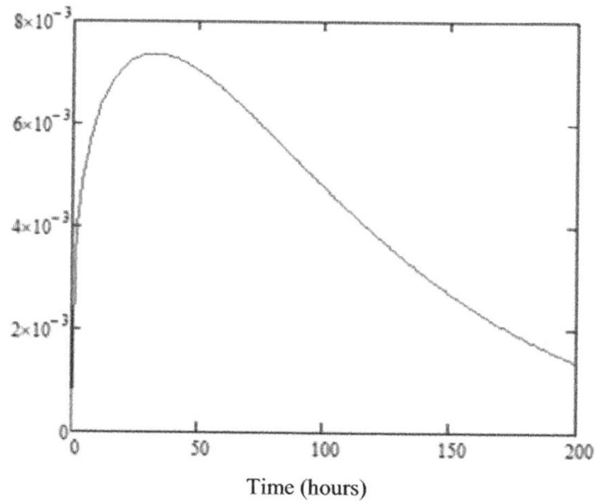

Figure 9.11. *Weibull probability density function for beta = 1.3*

9.6. Uptimes exponential distribution and repair times Weibull distribution

The normal distribution, with its symmetry around the mean and its domain of definition spreading for negative time values, is not always very suitable. Therefore, a Weibull distribution is often more realistic.

EXAMPLE 9.5.– Exponential distribution $\lambda = 0.005$ and Weibull distribution repair time with parameters $\eta = 40$ and $\beta = 1.3$. We obtain the results in Figure 9.12.

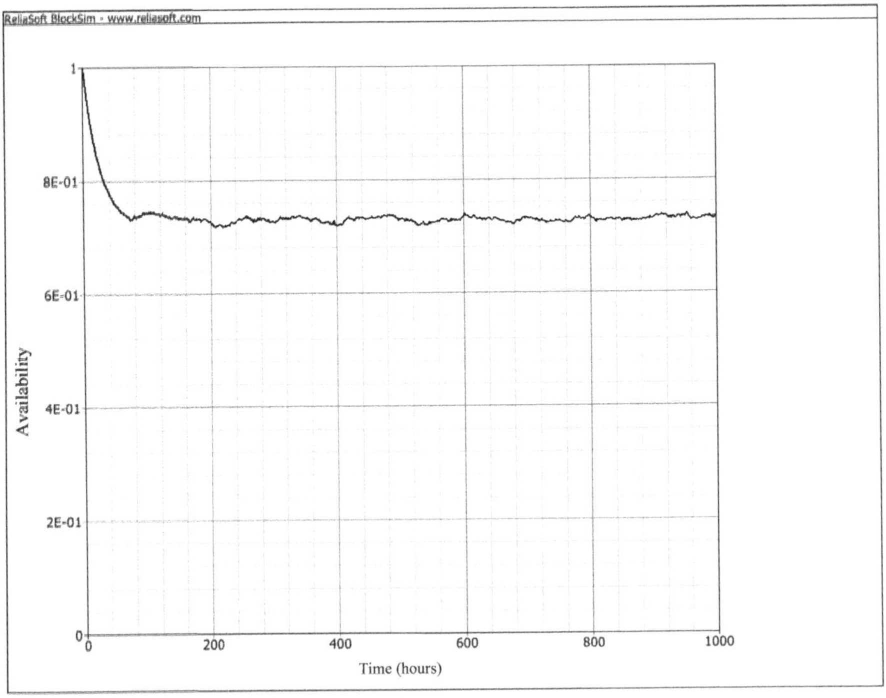

Figure 9.12. *Availability function for example 9.5*

Asymptotic availability is equal to ~ 0.73.

Availability/unavailability scheme

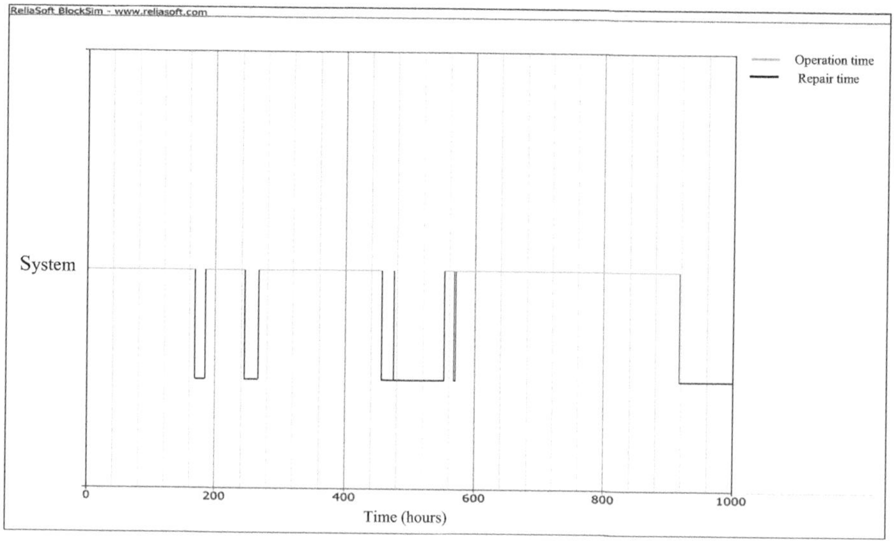

Figure 9.13. *Availability/unavailability scheme of example 9.5*

9.7. Serial system

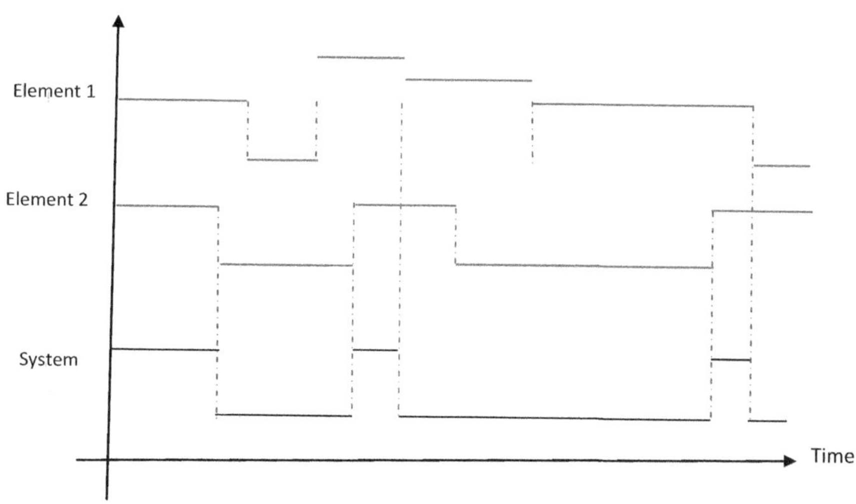

Figure 9.14. *Illustration of availability/unavailability scheme for a series system*

A serial system is available if and only if all of the components it contains are also available. We have the following illustration for a system with two elements.

For a serial system, the availability of a system composed of n elements is given by:

$$A_S(t) = \prod_{i=1}^{n} A_i(t)$$

For elements following an exponential distribution of parameter λi, we obtain:

$$A_S(t) = \prod_{i=1}^{n}\left(\frac{\mu_i}{\lambda_i+\mu_i} + \left(1 - \frac{\mu_i}{\lambda_i+\mu_i}\right).exp(-(\lambda_i + \mu_i).t)\right) \qquad [9.7]$$

The asymptotic availability is therefore equal to:

$$A_\infty = \prod_{i=1}^{n}\left(\frac{\mu_i}{\lambda_i+\mu_i}\right) \qquad [9.8]$$

EXAMPLE 9.6.– Exponential distribution λ1 = 0.001, μ1 = 0.1 and λ = 0.0012, μ2 = 0.15.

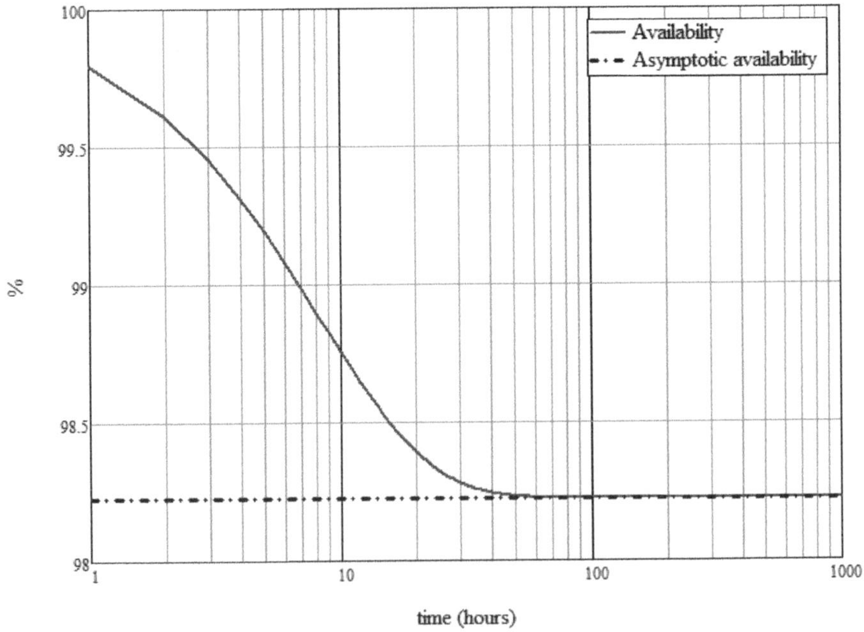

Figure 9.15. *Availability of an exponential series system*

Simulation

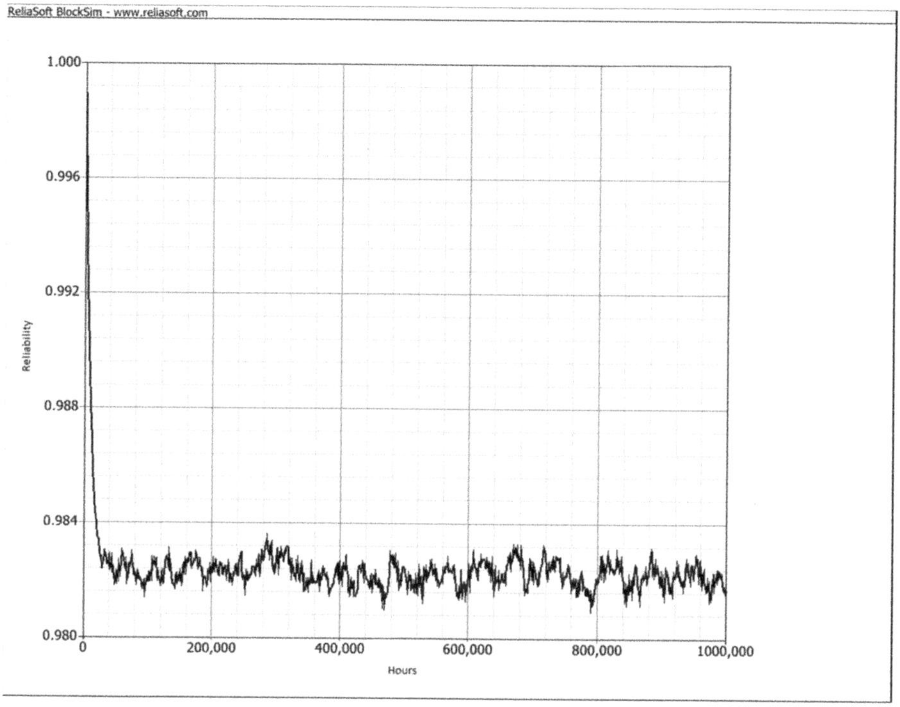

Figure 9.16. *Simulation of the availability of an exponential series system*

A total of 100,000 simulations are necessary to obtain a correct plot.

The asymptotic availability is obtained from equation [9.8] with n = 2, that is:

$$A_\infty = \left(\frac{\mu_1}{\lambda_1 + \mu_1}\right) \cdot \left(\frac{\mu_2}{\lambda_2 + \mu_2}\right) = \left(\frac{0.1}{0.001 + 0.1}\right) \cdot \left(\frac{0.15}{0.0012 + 0.15}\right) \approx 0.982$$

Availability/unavailability scheme

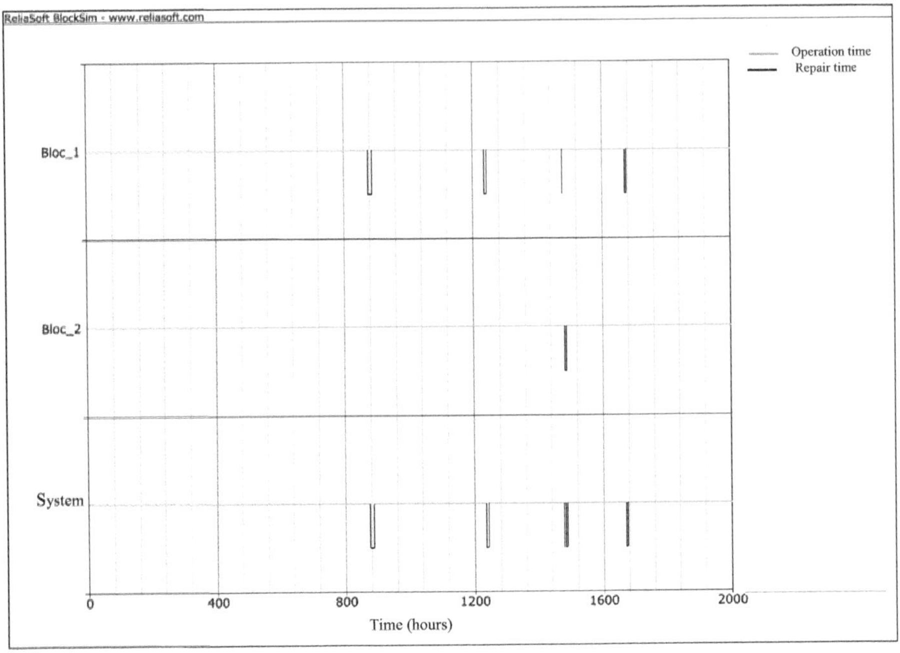

Figure 9.17. *Availability/unavailability scheme of series system with exponential distributions*

9.8. Parallel system

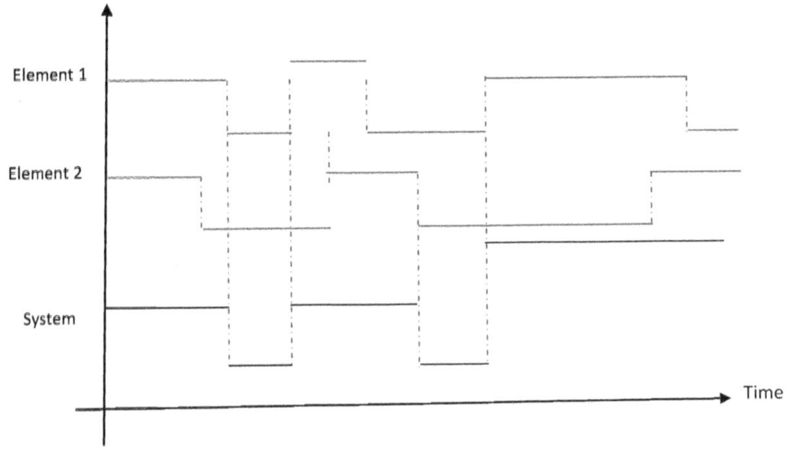

Figure 9.18. *Illustration of availability/unavailability scheme for a parallel system*

A parallel system is available if at least one of these elements is also available. We have the illustration shown in Figure 9.18.

The availability of a parallel system composed of n elements is given by:

$$A_S(t) = 1 - \prod_{i=1}^{n} \left(1 - A_i(t)\right)$$

In general, all of the elements are the same such that:

$$A_S(t) = 1 - \left(1 - A(t)\right)^n$$

For elements following an exponential distribution for uptime and repair times, we obtain:

$$A_S(t) = 1 - \left(1 - \frac{\mu}{\lambda+\mu} + \left(1 - \frac{\mu}{\lambda+\mu}\right) . exp(-(\lambda + \mu).t)\right)^n \qquad [9.9]$$

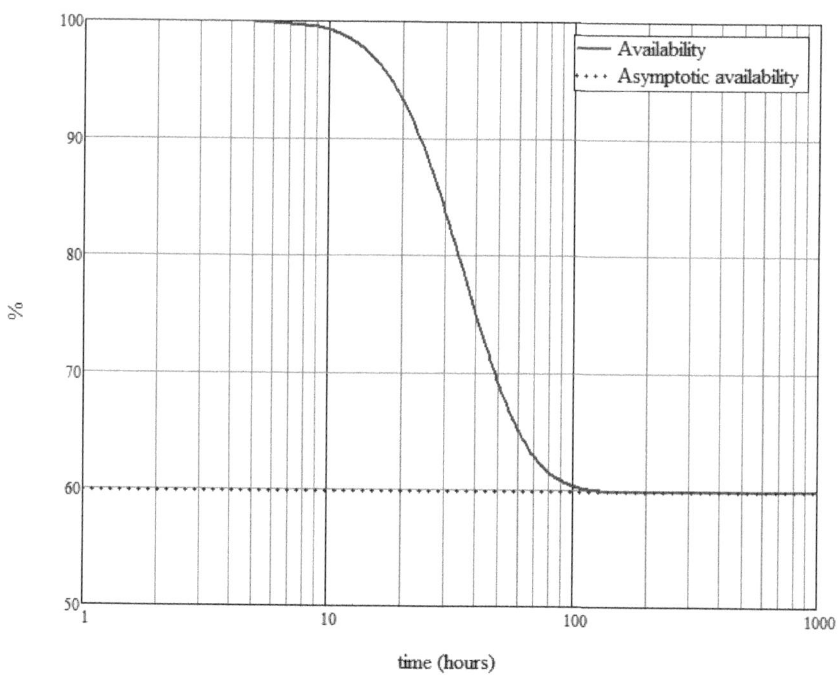

Figure 9.19. *Availability of an exponential parallel system*

The asymptotic availability is therefore equal (because the exponential tends to 0 when t tends to +∞).

$$A_\infty = 1 - \left(1 - \frac{\mu}{\lambda + \mu}\right)^n$$

That is:

$$A_\infty = 1 - \left(\frac{\lambda}{\lambda + \mu}\right)^n \qquad [9.10]$$

EXAMPLE 9.7.– $\lambda = 0.05$; $\mu = 0.01$; $n = 5$.

Simulation

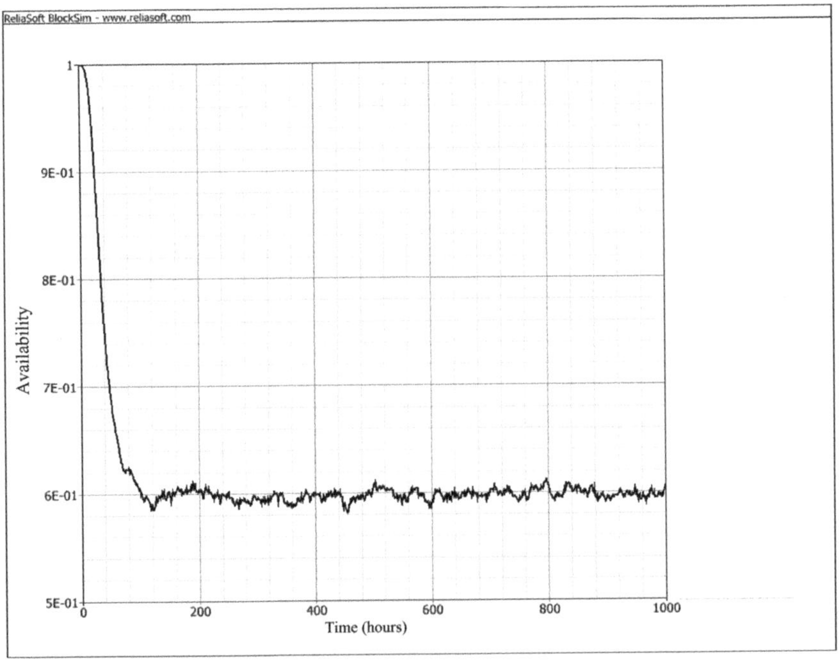

Figure 9.20. *Simulation of the availability of an exponential parallel system*

The point and asymptotic availabilities are obtained from equations [9.9] and [9.10], respectively.

Availability/unavailability scheme

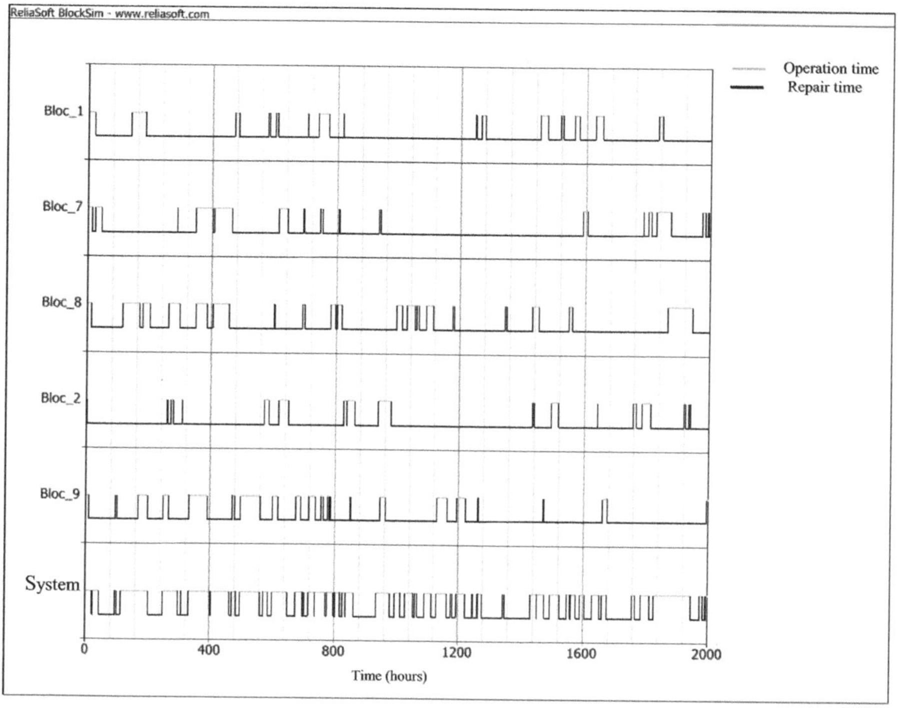

Figure 9.21. *Availability/unavailability scheme of parallel system of example 9.7*

9.9. k-out-of-n redundancy

For a k-out-of-n redundancy system, the availability A of a system composed of n elements, since all of the elements are identical, is given by:

$$A_S(t) = \sum_{i=k}^{n} C_n^i . A(t)^i . \left(1 - A(t)\right)^{n-i}$$

For elements following an exponential distribution for uptimes and repair times, we obtain by setting $\vartheta = \frac{\mu}{\lambda+\mu}$:

$$A_S(t) = \sum_{i=k}^{n} C_n^i . \left(\vartheta + (1-\vartheta) . exp(-(\lambda+\mu).t)\right)^i . \left(1 - \left(\vartheta + (1-\vartheta) . exp(-(\lambda+\mu).t)\right)\right)^{n-i} \quad [9.11]$$

The asymptotic availability is therefore equal to:

$$A_\infty = \sum_{i=k}^{n} C_n^i \cdot \left(\frac{\mu}{\lambda+\mu}\right)^i \cdot \left(\frac{\lambda}{\lambda+\mu}\right)^{n-i} \quad [9.12]$$

EXAMPLE 9.8.– $\lambda = 0.01$; $\mu = 0.05$; $k = 2$; $n = 4$.

Asymptotic availability is equal to ~ 0.997.

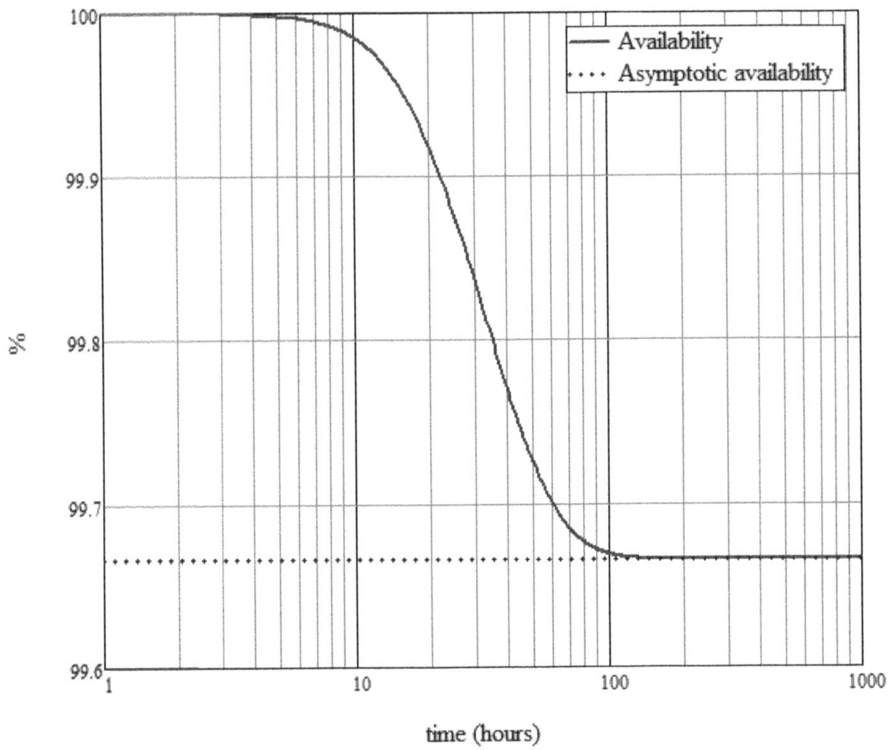

Figure 9.22. *Availability of an exponential k/n system*

Simulation

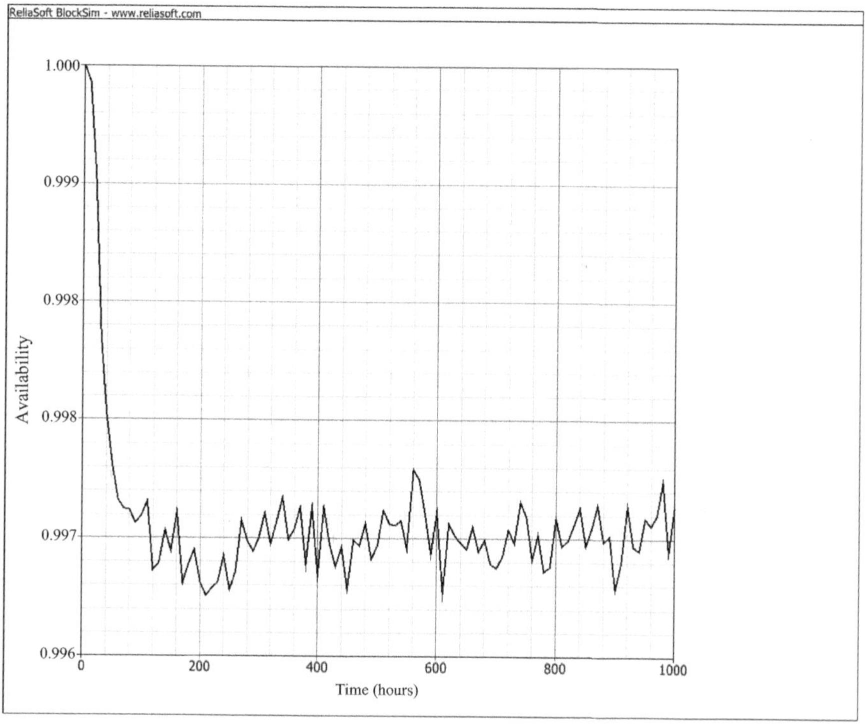

Figure 9.23. *Simulation of the availability of an exponential k/n system*

Availability/unavailability scheme

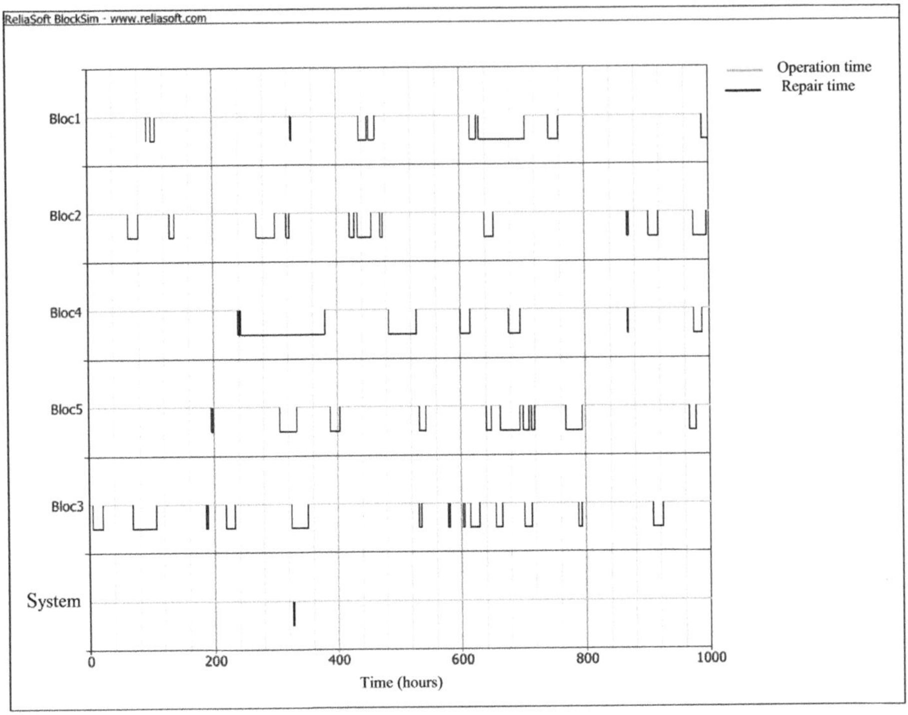

Figure 9.24. *Availability/unavailability scheme of example 9.8*

9.10. Series/parallel system

These systems are made up of "p" subsets (S/E) in series. Each subset consists of a number of elements assembled in parallel and unique thereto. Therefore, for subset "i", the number of elements will be "ni".

These systems are illustrated in Figure 3.3.

The system will then be available if all the subsets are also available, that is:

$$A_S(t) = \prod_{i=1}^{p} A_{S/E_i}(t)$$

If we denote by $A_{i,j}$ the availability of the "*i*th" element of the subset "j", we can write that:

$$A_{S/E_i} = 1 - \prod_{j=1}^{n_i} \left(1 - A_{i,j}(t)\right)$$

from which it is derived that:

$$A_S(t) = \prod_{i=1}^{p} \left(1 - \prod_{j=1}^{n_i} \left(1 - A_{i,j}(t)\right)\right)$$

Generally, since we are dealing with a parallel setup, the elements of each subset are identical so that the availability of the system can finally be written as:

$$A_S(t) = \prod_{i=1}^{p} \left(1 - \left(1 - A_i(t)\right)^{n_i}\right) \quad [9.13]$$

REMARK.–

If the values of $A_{i,j}$ tend to 1, then the values of $(1 - A_{i,j})$ tend to 0. Therefore, the product $\prod_{i=1}^{n_j} \left(1 - A_{i,j}(t)\right)$ also tends to 0. Consequently, $1 - \prod_{i=1}^{n_j} \left(1 - A_{i,j}(t)\right)$ tends to be 1. This means that the availability of the system tends to 0 since we take the product of values less than 1. This type of system architecture is highly inadequate from an availability point of view and is seldom found in industry.

9.11. Parallel/serial system

These systems are made up of "p" subsets (S/E) in parallel. Each subset consists of a number of elements unique thereto.

Therefore, for subset "i", the number of elements will be "p_i". These systems are illustrated in Figure 3.36.

The system will then be unavailable if all of the subsets are also unavailable, namely:

$$\overline{A_S}(t) = \prod_{i=1}^{n} \overline{A_{S/E_i}}(t) = \prod_{i=1}^{n} \left(1 - A_{S/E_i}\right)$$

As a result, the system availability is written as:

$$A_S(t) = 1 - \prod_{i=1}^{n} \left(1 - A_{S/E_i}\right)$$

If we denote by $A_{i,j}$ the availability of the "*i*th" element of the subset "j", we can write that:

$$A_{S/E_i} = \prod_{j=1}^{p_i} A_{i,j}(t)$$

from which it is derived that:

$$A_S(t) = 1 - \prod_{i=1}^{n} \left(1 - \prod_{j=1}^{p_i} A_{i,j}(t)\right)$$

Typically, the elements of each subset are identical so that system availability can ultimately be written as:

$$A_S(t) = 1 - \left(1 - \prod_{j=1}^{p} A_j(t)\right)^n \quad [9.14]$$

where p is the number of elements in series in a subset.

REMARK.–

If the values $A_{i,j}$ tend to 1, then the product $\prod_{i=1}^{n_j} A_{i,j}(t)$ tends to 0.

Consequently, $1 - \prod_{i=1}^{n_j} \left(1 - A_{i,j}(t)\right)$ tends to 1. Therefore, the product $\prod_{i=1}^{n} \left(1 - \prod_{j=1}^{p_i} A_{i,j}(t)\right)$ tends to 0, and therefore, the system availability tends to 1.

This type of system architecture is therefore highly suitable from an availability point of view and is often found in industry (as for civil avionics with "Pilot/Copilot" redundancy).

9.12. Energy conversion

Reconsider the example of Figure 3.43 in Chapter 3. According to equation [9.13], we have:

$$A_S(t) = \prod_{j=1}^{1}\left(1 - \left(1 - A_j(t)\right)^{n_j}\right)$$

with $n_1 = 1$ and $n_2 = 2$. Hence:

$$A_S(t) = \left(1 - \left(1 - A_1(t)\right)\right).\left(1 - \left(1 - A_2(t)\right)^2\right)$$

that is,

$$A_S(t) = A_1(t).A_2(t).\left(2 - A_2(t)\right)$$

Let us assume that the electronics are modeled by an exponential distribution $\lambda = 10^{-5}$, whereas the fans are modeled by a Weibull distribution with parameters $\eta = 4{,}000$ and $\beta = 2.4$. We obtain for a repair time modeled by a uniform distribution of parameters [200; 1,000]:

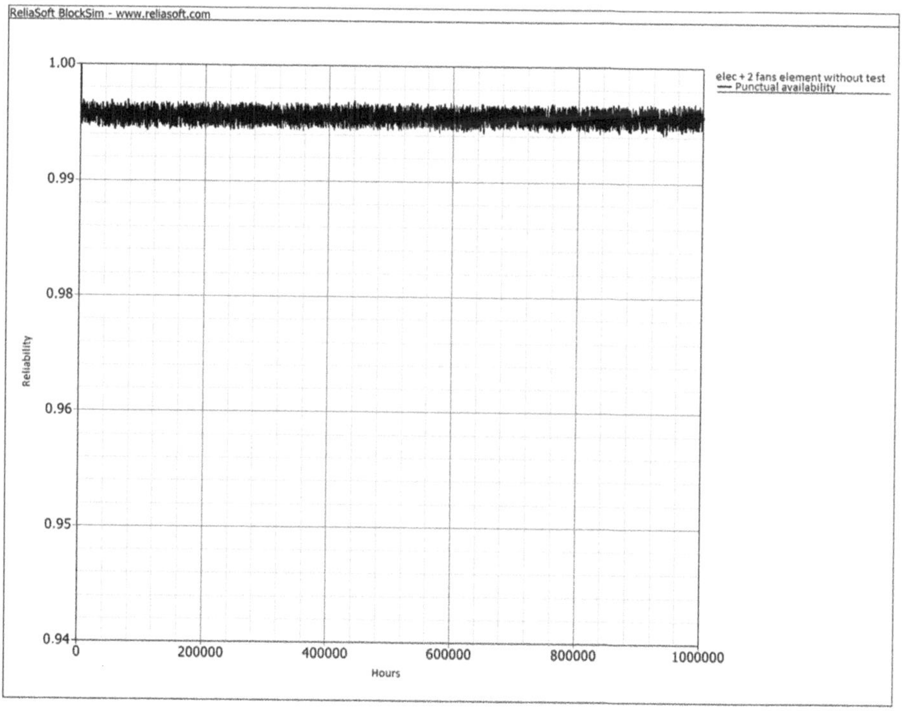

Figure 9.25. *Simulation of the availability of conversion energy system*

Availability/unavailability scheme

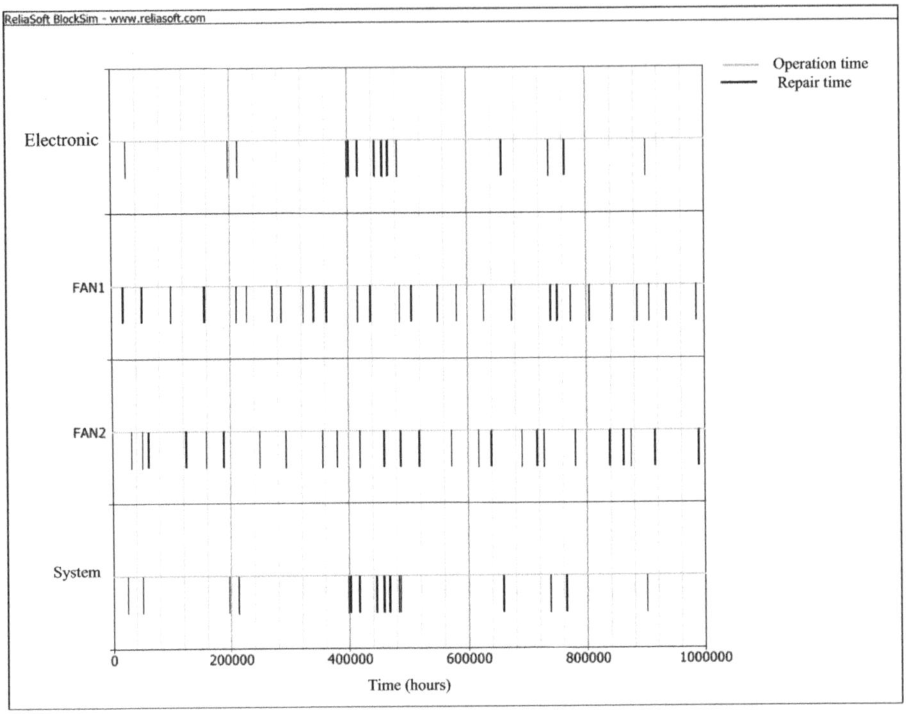

Figure 9.26. *Availability/unavailability scheme of previous example*

Avionics system: electronics exponential $\lambda = 0.001$; IHM interface Weibull $\eta = 400$; $\beta = 4.7$; repair times uniform distribution [50; 100].

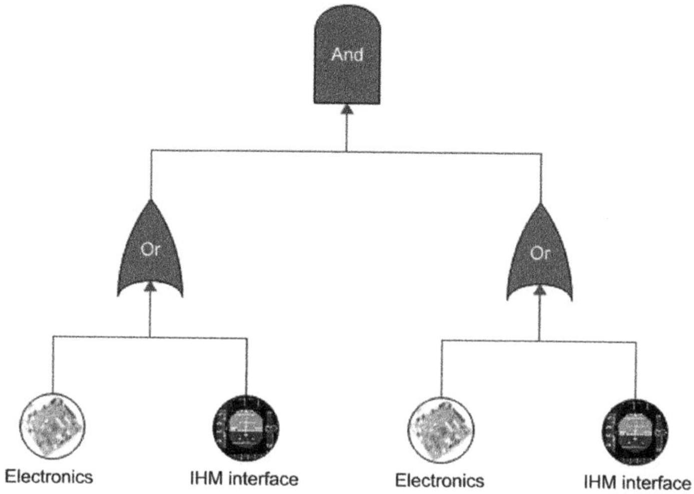

Figure 9.27. *System avionic block diagram*

Element-level renewal

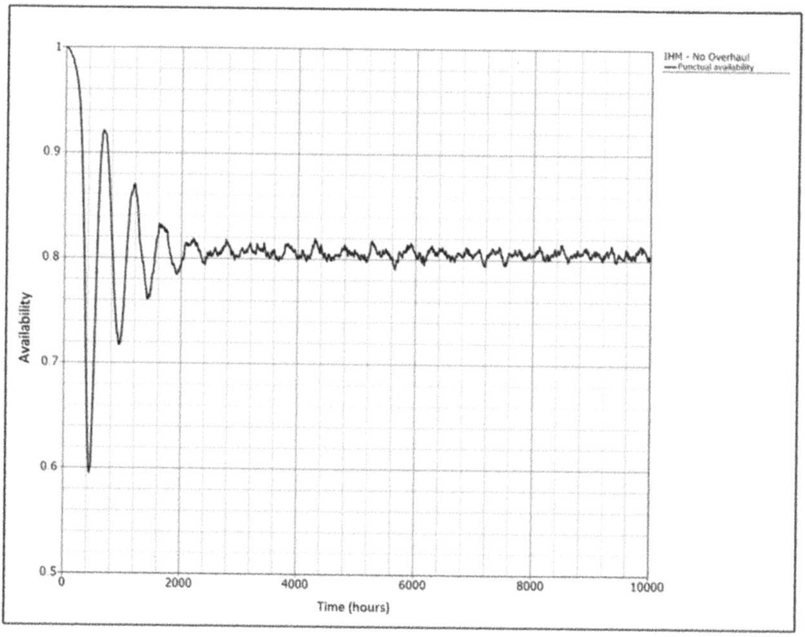

Figure 9.28. *Simulation of the availability of the avionic system*

Availability/unavailability scheme

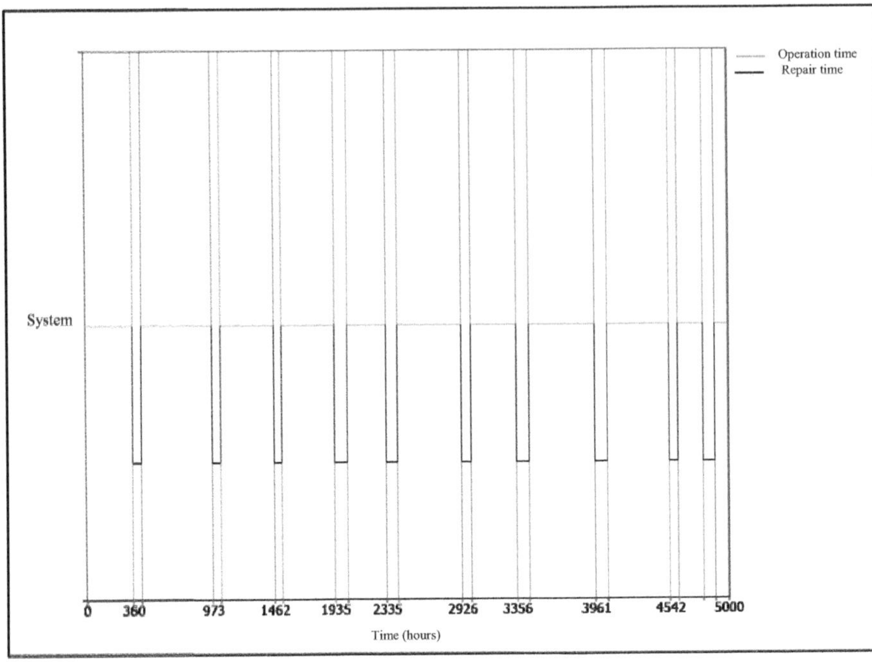

Figure 9.29. *Availability/unavailability scheme of avionic system at element level*

System-level renewal

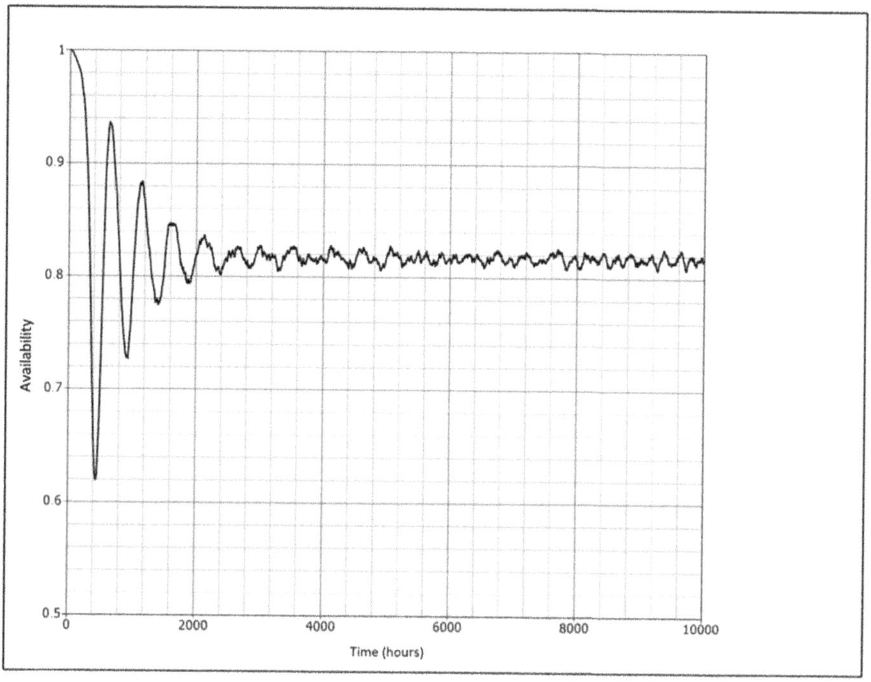

Figure 9.30. *Simulation of the availability of the avionic system at system level*

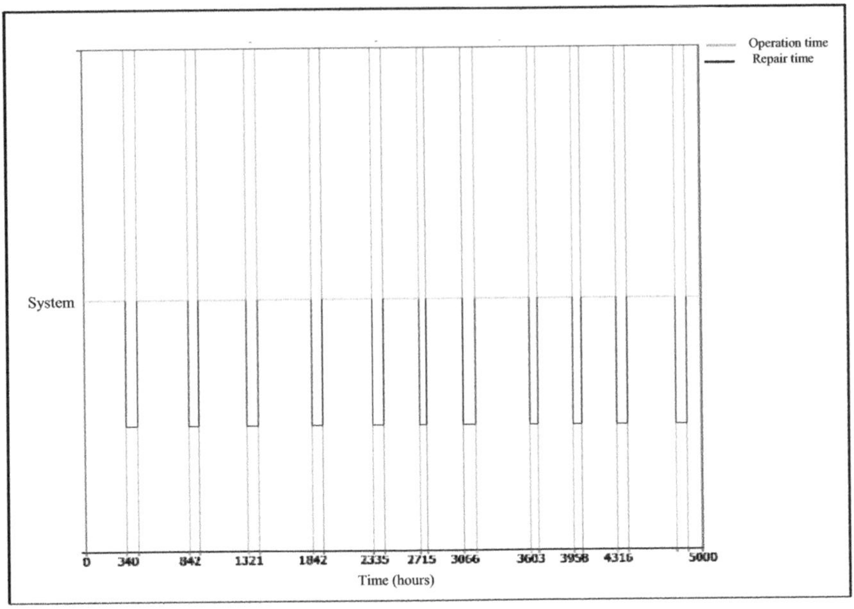

Figure 9.31. *Availability/unavailability scheme of the avionic system with maintenance at system level*

9.13. Summary

Generally speaking, we do not have explicit solutions for availability. Although not truly reflecting the industrial context, only a few cases give explicit results:

– When failure and repair times follow an exponential distribution, we have an explicit solution of availability.

– When failure times follow an exponential distribution and repair times are constant, we have an explicit solution of availability.

– When failure times follow an exponential distribution and information about the repair times is scarce, these can be modeled by a uniform distribution.

– When failure times follow an exponential distribution and there is a certain amount of information on the repair times, these can be modeled by a normal distribution.

– If the repair times are not symmetrical with respect to the mean value, other probability distributions can be considered, such as the Weibull distribution or even the noncentral beta distribution proposed in Chapter 7, when the skewness of the repair times is very pronounced.

Part 4

Safety

10

FMEA Concurrent Failure Mechanisms

A component typically has multiple failure mechanisms. If we assume the independence of these mechanisms, we can regard it as a serial system since it suffices that a failure is observed for a given failure mechanism for the component to fail. We should remember that a mechanism is the cause of a failure while the mode is the consequence, that is, the state of the component that will be observed in practice. The relationship between mechanism and mode is not bijective because, for example, two different mechanisms can generate the same mode.

These failure mechanisms are then called "concurrent" or "competitive".

Knowledge of the failure mechanisms of a component is extremely important because it will lead to:

– Identifying potential aging mechanisms. Among the three possible types of failure, aging mechanisms are the most feared because they impact all of the products. Therefore, although they have virtually no information from component manufacturers about this matter, designers must ensure that these mechanisms will not be observed under operational conditions. To eliminate any doubts, they may be required to perform specific aging tests.

– Building the failure mode and effects analysis (FMEA), which is a crucial step and very often a customer requirement. It is, in principle, carried out at the component level, which requires much work. It is also a difficult analysis because component manufacturers do not usually provide any information on this topic. In addition, the increasing complexification of digital components such as microprocessors, memory, ASICs, etc., has made this task increasingly complex.

– Developing failure trees of feared events whose probability of occurrence is very often required by the customer.

For a color version of all of the figures in this chapter, see www.iste.co.uk/bayle/montecarlo.zip.

In all of the cases previously mentioned, not only the nature of the failure mechanism but also the nature of the failure mode (short circuit, performance drift, etc.) to which it leads, must be identified. The component manufacturer sometimes provides this information.

For example, we can mention Nichicon for electrolytic capacitors (Nichicon Corporation 2000), as illustrated in Figure 10.1.

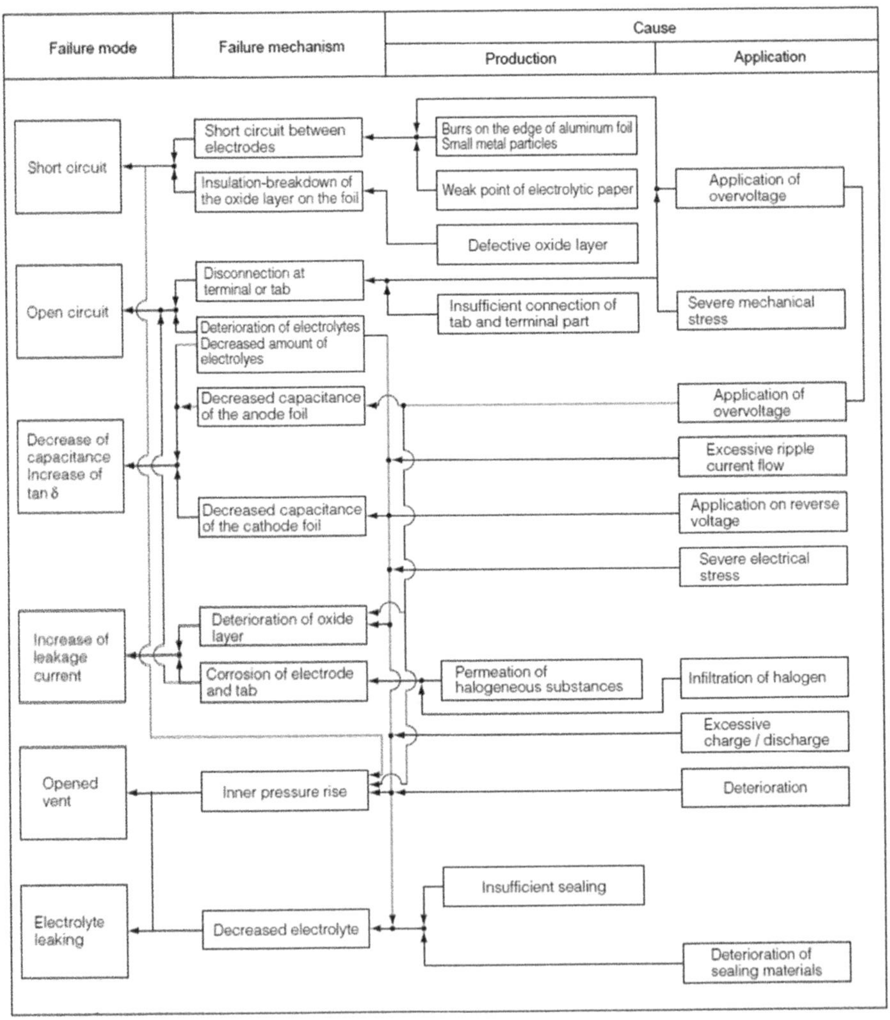

Figure 10.1. *Mechanisms and failure modes in an electrolytic capacitor from Nichicon*

In addition, the quantification of each of the failure modes is necessary to estimate the failure rate of the effect at the product level.

For this purpose, it is possible to use certain standards that give the percentage of occurrence of each failure mode of a given component (Quanterion Solutions Incorporated 2016), as illustrated in Table 10.1.

Device Type	Failure Mode	Failure Mode Probability (α)
Coil	Short	.42
	Open	.42
	Change in Value	.16
Computer System	Hardware Failure	.57
	Software Failure	.43
Connector/Connection	Open	.61
	Poor Contact/Intermittent	.23
	Short	.16
Controller, Electromechanical	Erroneous Output	.75
	Loss of Control	.25
Counter Assembly	Inaccurate Count	.91
	Seized	.09
Crystal, Quartz	Open	.89
	No Oscillation	.11
Diode, General	Short	.49
	Open	.36
	Parameter Change	.15
Diode, Rectifier	Short	.51
	Open	.29
	Parameter Change	.20
Diode, SCR	Short	.98
	Open	.02
Diode, Small Signal	Parameter Change	.58
	Open	.24
	Short	.18
Diode, Thyristor	Failed Off	.45
	Short	.40
	Open	.10
	Failed On	.05
Diode, Triac	Failed Off	.90
	Failed On	.10
Diode, Zener, Voltage Reference	Parameter Change	.69
	Open	.18
	Short	.13
Diode, Zener, Voltage Regulator	Open	.45
	Parameter Change	.35
	Short	.20
Electric Motor, AC	Winding Failure	.31
	Bearing Failure	.28
	Fails to Run, After Start	.23
	Fails to Start	.18

Table 10.1. *Example of the distribution of failure modes in 1991 FMD*

Alternatively, for an integrated circuit from the Texas Instrument (Texas Instrument 2021), see Figure 10.2.

Automotive 3-V to 36-V, 4-A, Low-Noise Synchronous Step-Down Converter

Functional Block Diagram

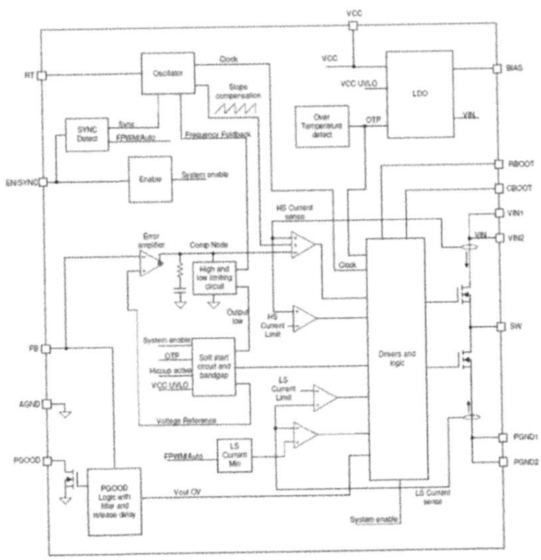

Failure Rate Mission Profile (1)	Per 10^9 Hours (FIT)
Total FIT Rate	16
Die FIT Rate	8
Package FIT Rate	8

Failure Modes	Failure Mode Distribution (%)
SW No output	45%
SW output not in specification – voltage or timing	40%
SW power FET stuck on	5%
PGOOD false trip, fails to trip	5%
Short circuit any two pins	5%

Figure 10.2. *Failure mode distribution for the Texas Instrument*

If we reconsider the three possible types of failure, to analyze the distribution of failure modes in a component, we can eliminate "infant mortality" failures because they are small in number and potentially filtered by burn-in.

In addition, these mechanisms are difficult to understand because they essentially depend on manufacturing defects specific to a manufacturer, which may vary over time and differ from one manufacturer to another.

All that exists in the universe is sooner or later affected by an aging mechanism, and this is the case for all components. Even a 10 kΩ resistor through which a current of 1 mA flows will be subject to failure mechanisms. These will generally not be observed during operation because the operation time (service life) will be much shorter than their lifespan. Nonetheless, for certain components (mechanical, electromechanical, optical, etc.) known as "limited lifespan" components, and depending on operational conditions, aging mechanisms may be observed.

This then raises three problems:

– the aging mechanism model does not always consider the failure mode distributions discussed previously;

– reliability information, when specified, as for the Texas Instrument, is not applicable for aging mechanisms because it is time independent;

– the impact of maintenance is also not explained.

The objective of this chapter is to show how these distributions should be correctly estimated. Finally, failures with a catalectic origin are time independent because they are sudden, since they are induced by an event extrinsic to the component. They are generally modeled by an exponential distribution whose main particularity is to have a constant failure rate. If the products are maintained under operational conditions, they can then be modeled using a homogeneous Poisson process (HPP), which is also a renewal process.

10.1. Maintenance-free industrial applications

Let "m" be the number of failure mechanisms of the component under consideration. Under the assumption that they are independent of one another, the component reliability function is given by:

$$R(t) = \prod_{i=1}^{m} R_i(t)$$

It is thus shown that its failure rate is given by (Appendix 5):

$$h(t) = \sum_{i=1}^{m} h_i(t)$$

The distribution of the "*i*th" failure mechanism ξ_i can then be defined by:

$$\xi_i(t) = \frac{h_i(t)}{h(t)} = \frac{h_i(t)}{\sum_{j=1}^{m} h_j(t)} \quad [10.1]$$

10.1.1. *Exponential distribution*

Here, the failure rates are constant, which thereby simply yields:

$$\xi_i(t) = \frac{h_i}{\sum_{j=1}^{m} h_j} \quad [10.2]$$

The distribution of catalectic failure mechanisms is therefore not time dependent.

EXAMPLE 10.1.– Mechanism 1: λ1 = 0.01 and mechanism 2: λ1 = 0.012; therefore:

$$\zeta 1 = \frac{\lambda 1}{\lambda 1+\lambda 2} = \frac{0.01}{0.01+0.012} = 45.455\% \qquad \zeta = \frac{\lambda 2}{\lambda 1+\lambda 2} = \frac{0.012}{0.01+0.012} = 54.545\%$$

10.1.2. *Weibull distribution*

Here, from a theoretical point of view, the following is obtained:

$$\xi_i(t) = \frac{\left(\frac{\beta_i}{\eta_i}\right)\cdot\left(\frac{t}{\eta_i}\right)^{\beta_i-1}}{\sum_{j=1}^{m}\left(\frac{\beta_j}{\eta_j}\right)\cdot\left(\frac{t}{\eta_j}\right)^{\beta_j-1}} \quad [10.3]$$

The distribution of aging failure mechanisms obviously depends on time.

EXAMPLE 10.2.– Weibull η1 = 1,000, β1 = 2 and η2 = 1,200, β2 = 7.

The theoretical calculation gives in this case:

$$\xi_1(t) = \frac{\left(\frac{\beta_1}{\eta_1}\right)\cdot\left(\frac{t}{\eta_1}\right)^{\beta_1-1}}{\sum_{j=1}^{2}\left(\frac{\beta_j}{\eta_j}\right)\cdot\left(\frac{t}{\eta_j}\right)^{\beta_j-1}} \text{ and } \xi_2(t) = 1 - \xi_1(t)$$

We obtain the results in Figure 10.3.

At the beginning, Mechanism 1 will essentially be observed because its scale parameter η is smaller and the failure times are more scattered because of its smaller shape parameter.

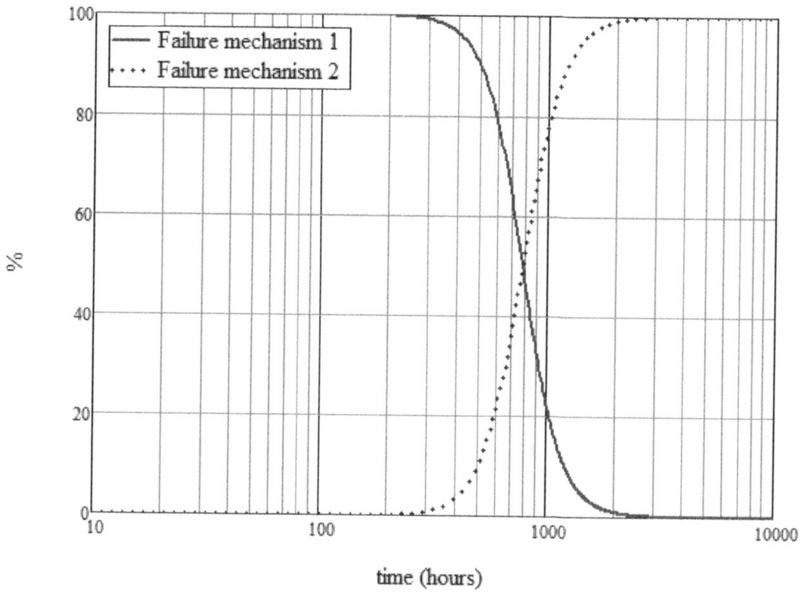

Figure 10.3. *Distribution of failure modes with two Weibull distributions without maintenance by computation*

For example, for applications without maintenance, analytical solutions are sufficient for estimating the distributions of the failure mechanisms of a component. In the next section, with the exception of the exponential distribution, we shall see that this is no longer the case and that Monte Carlo simulations make estimating these distributions possible.

10.2. Industrial applications with maintenance

Maintenance is assumed to be ideal (negligible repair times compared with times of proper operation). Since a component is not repaired for essentially technical reasons, maintenance is thus performed at the component level and not at the failure mechanism level. We therefore have a renewal process at the component level, and as such, we come about the notion of Rocof.

Remember that this is defined by:

$$Rocof(t) = \sum_{k=1}^{+\infty} f^{<k>}(t)$$

The component probability of failure is given by:

$$F(t) = 1 - R(t) = 1 - \prod_{i=1}^{m} R_i(t)$$

As such, the probability density is given by (Appendix 3):

$$f(t) = \sum_{i=1}^{m} \left(f_i(t). \prod_{\substack{j=1 \\ j \neq i}}^{m} R_j(t) \right)$$

10.2.1. *Exponential distributions with parameter λ_i*

We have here: $f_i(t) = \lambda_i . exp(-\lambda_i . t). \beth_{t\geq0}$. In this case, the component probability density is given by:

$$\mathrm{f}(t) = \sum_{i=1}^{m} \lambda_i . exp\left(- \sum_{i=1}^{m} \lambda_i . t \right) . \beth_{t\geq0}$$

The Rocof is therefore equal to (Bayle 2019):

$Rocof(t) = \sum_{i=1}^{m} \lambda_i$ and thereby $N(t) = \sum_{i=1}^{m} \lambda_i . t$

There is thus a superposition of the "m" renewal process, which, for the exponential distribution, still represents a renewal process, as illustrated below for two failure mechanisms.

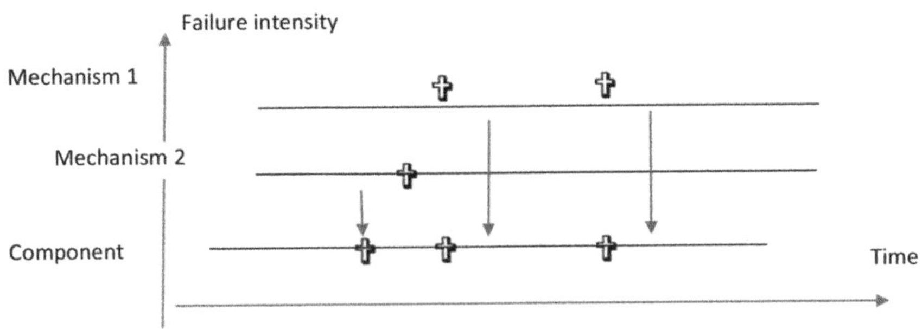

Figure 10.4. *Illustration of the failure intensity for a series system*

We can therefore estimate the distribution of failure mechanisms by:

$$\xi_i(t) = \frac{N_i(t)}{\sum_{j=1}^{m} N_j(t)} = \frac{\lambda_i . t}{\sum_{j=1}^{m} \lambda_j . t}$$

or still

$$\xi_i(t) = \frac{\lambda_i}{\sum_{j=1}^{m} \lambda_j} \qquad [10.4]$$

The distribution of failure mechanisms is not time dependent. This result makes sense because the exponential distribution is memoryless.

As such, maintenance has no impact on the distribution of failure mechanisms.

EXAMPLE 10.3.– Exponential distributions $\lambda 1 = 10^{-2}$ and $\lambda 2 = 1.5\ 10^{-2}$.

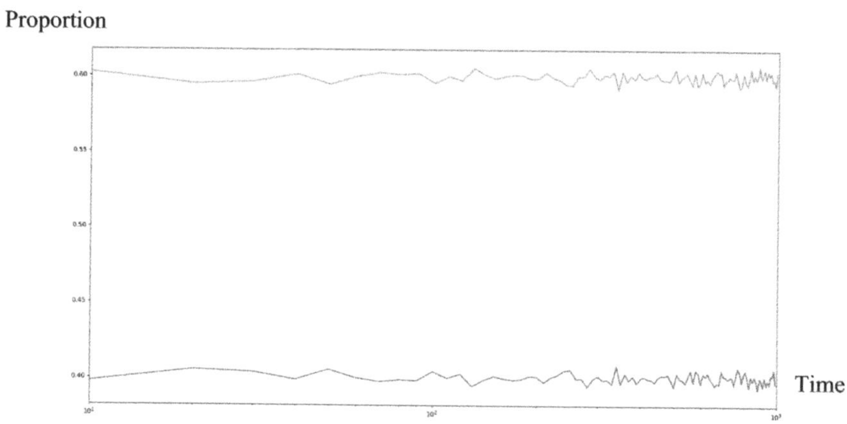

Figure 10.5. *Simulation of failure mechanism repartition versus time in the case of exponential distributions*

10.2.2. *Weibull distributions*

Within the context of aging phenomena, since the failure rate is increasing, we no longer have a renewal process at the component level, as illustrated in Figure 10.6.

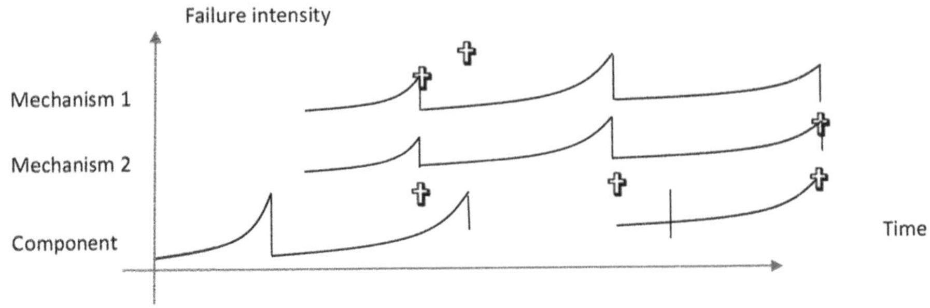

Figure 10.6. *Simulation of failure mechanism repartition versus time in the case of Weibull distributions*

Indeed, by replacing the component with a new component, we reset all of the failure mechanisms, even if they have not been observed. There are thus no explicit solutions for estimating the distribution of the component failure mechanisms, and we will thus utilize Monte Carlo simulations to obtain an estimation thereof.

EXAMPLE 10.4.– Normal distributions: μ1 = 1,000; σ1 = 100 and μ2 = 1,400; σ2 = 500.

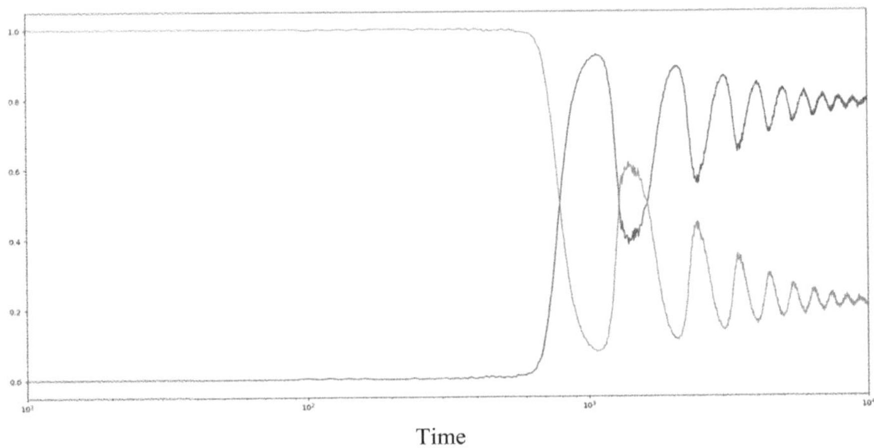

Figure 10.7. *Distribution of failure mechanisms with ideal maintenance normal distributions*

Initially, Mechanism 1 is predominant because of its smaller scale parameter, μ1. The oscillations are important because of the low coefficient of variation of mechanism 1 (CV1 = 100/1,000 = 10%).

EXAMPLE 10.5.– Weibull distributions: η1 = 100, β1 = 1.5 and η2 = 120 β = 5.

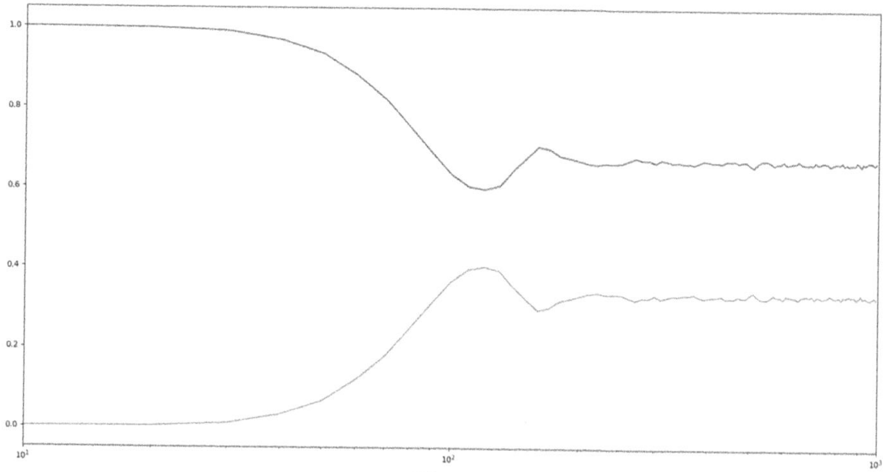

Figure 10.8. *Distribution of fault mechanisms with ideal maintenance Weibull distributions*

10.3. Consideration of physical contributions

The level of physical contribution usually has a significant impact on component reliability. Therefore, we reuse the temperature example and make the assumption of an AFT (accelerated failure time: an increase in temperature accelerates time) model. We also make the traditional assumption of modeling the effect of temperature on reliability using an Arrhenius law.

10.3.1. *Exponential law with no maintenance*

In this case, the failure rate can be written as:

$$h(\theta) = \mathrm{C}.\exp\left(-\frac{Ea}{Kb.\theta}\right)$$

If we recover equation [10.1], we therefore obtain:

$$\xi_i(\theta) = \frac{h(\theta)}{\sum_{j=1}^{m} h_j(\theta)}$$

That is, with the Arrhenius law:

$$\xi_i(\theta) = \frac{C_i.\exp\left(-\frac{Ea_i}{Kb.\theta}\right)}{\sum_{j=1}^{m} C_j.\exp\left(-\frac{Ea_j}{Kb.\theta}\right)} \quad [10.5]$$

EXAMPLE 10.6.–

Three failure mechanisms with the following characteristics are considered: Mechanism 1: C1 = 2.933 × 10^{-7}; Ea1 = –0.2 eV; Mechanism 2: C2 = 40; Ea2 = 0.3 eV; and Mechanism 3: C3 = 1.463 107; Ea3 = 0.7 eV.

Since the distribution of failure mechanisms is time independent for exponential distributions, varying the temperature between –40°C and +125°C might be interesting. We obtain the results in Figure 10.9.

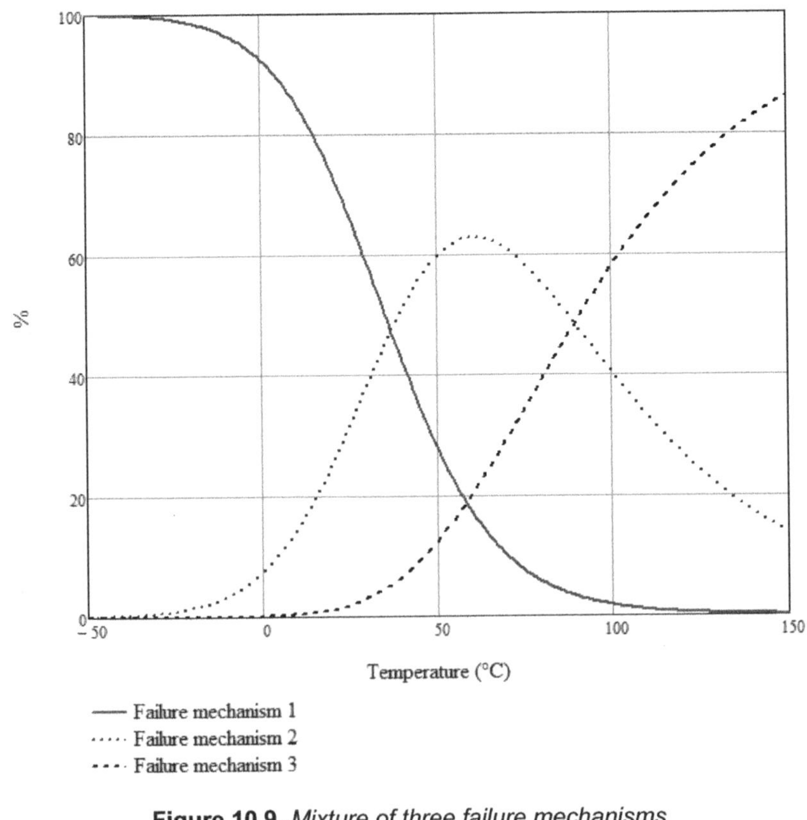

Figure 10.9. *Mixture of three failure mechanisms according to temperature exponential laws*

When the temperature is negative, only failure Mechanism 1 will be observed because of its negative activation energy. When the temperature is high (> 120°C), Mechanism 3 will be observed mainly because of its higher activation energy. The failure rate of the component according to temperature is shown in Figure 10.10.

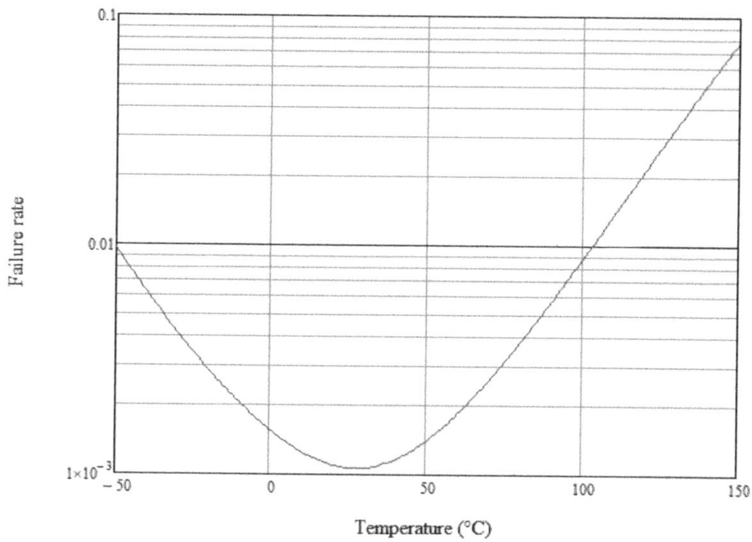

Figure 10.10. *Component failure rate as a function of the temperature exponential laws*

10.3.2. *Weibull distribution with no maintenance*

The distribution of failure mechanisms depends on time but also on temperature.

From the equation, we obtain:

$$\xi_i(t,\theta) = \frac{\left(\dfrac{\beta_i}{C_i.\exp\left(\frac{Ea_i}{Kb.\theta}\right)}\right)\cdot\left(\dfrac{t}{C_i.\exp\left(\frac{Ea_i}{Kb.\theta}\right)}\right)^{\beta_i-1}}{\sum_{j=1}^{m}\left(\dfrac{\beta_j}{C_j.\exp\left(\frac{Ea_j}{Kb.\theta}\right)}\right)\cdot\left(\dfrac{t}{C_j.\exp\left(\frac{Ea_j}{Kb.\theta}\right)}\right)^{\beta_j-1}} \qquad [10.6]$$

EXAMPLE 10.7.– Consider two failure mechanisms according to two Weibull distributions: C1 = 8.83 × 10^{-5}; β1 = 1.5; Ea1 = 0.4 eV and C2 = 5.18 × 10^{-10}; β2 = 5; Ea2 = 0.7eV.

The distribution of failure mechanisms is therefore time and temperature dependent. We obtain the following distributions for the failure mechanisms:

$$\xi_1(t,\theta)=\frac{\left(\frac{\beta_1}{C_1.\exp\left(\frac{Ea_1}{Kb.\theta}\right)}\right)\cdot\left(\frac{t}{C_1.\exp\left(\frac{Ea_1}{Kb.\theta}\right)}\right)^{\beta_1-1}}{\left(\frac{\beta_1}{C_1.\exp\left(\frac{Ea_1}{Kb.\theta}\right)}\right)\cdot\left(\frac{t}{C_1.\exp\left(\frac{Ea_1}{Kb.\theta}\right)}\right)^{\beta_1-1}+\left(\frac{\beta_2}{C_2.\exp\left(\frac{Ea_2}{Kb.\theta}\right)}\right)\cdot\left(\frac{t}{C_2.\exp\left(\frac{Ea_2}{Kb.\theta}\right)}\right)^{\beta_2-1}}$$

$$\xi_2(t,\theta)=\frac{\left(\frac{\beta_2}{C_2.\exp\left(\frac{Ea_2}{Kb.\theta}\right)}\right)\cdot\left(\frac{t}{C_2.\exp\left(\frac{Ea_2}{Kb.\theta}\right)}\right)^{\beta_2-1}}{\left(\frac{\beta_1}{C_1.\exp\left(\frac{Ea_1}{Kb.\theta}\right)}\right)\cdot\left(\frac{t}{C_1.\exp\left(\frac{Ea_1}{Kb.\theta}\right)}\right)^{\beta_1-1}+\left(\frac{\beta_2}{C_2.\exp\left(\frac{Ea_2}{Kb.\theta}\right)}\right)\cdot\left(\frac{t}{C_2.\exp\left(\frac{Ea_2}{Kb.\theta}\right)}\right)^{\beta_2-1}}$$

Figure 10.11 shows the distribution of the failure mechanisms according to time for three different temperatures (30°C, 50°C and 70°C).

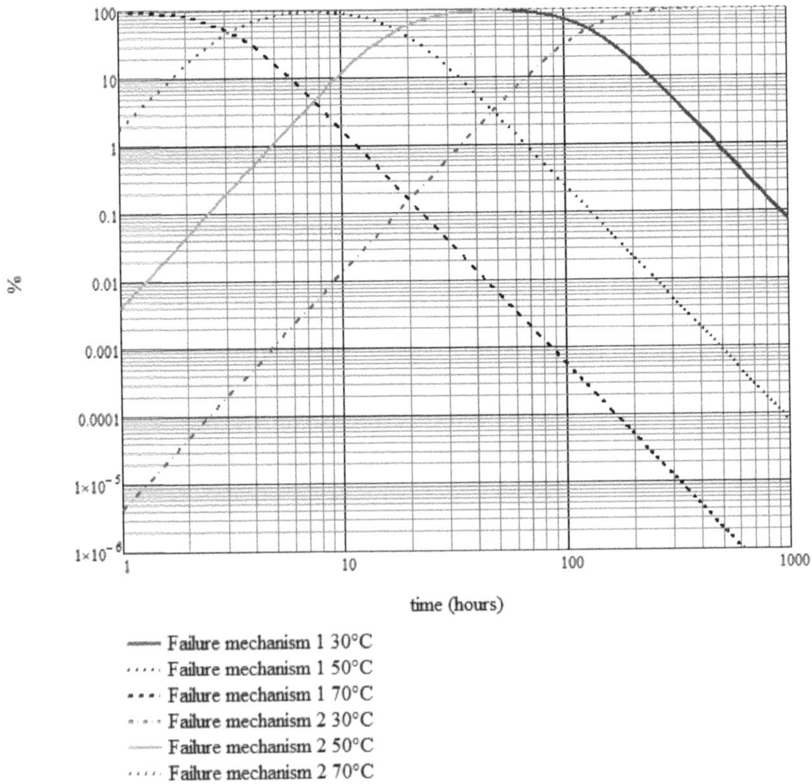

Figure 10.11. *Distributions of failure mechanisms for two Weibull distributions*

Clearly, the two failure mechanisms occur earlier at higher temperatures. At 30°C, although its scale parameter η is relatively high, it appears that its shape parameter β results in more scattered failure times.

At 50°C, the activation energy of the second failure mechanism increases, and its scale parameter η becomes much smaller than that of the first failure mechanism. Therefore, the first failure mechanism is predominant until t = 10 h, after which the second failure mechanism takes over. For a temperature of 70°C, the second failure mechanism is now predominant for the same reason as for 50°C.

10.3.3. *Exponential distribution with maintenance*

As we previously observed, the same expression as without maintenance comes across, from which:

$$\xi_i(\theta) = \frac{C_i.\exp\left(-\frac{Ea_i}{Kb.\theta}\right)}{\sum_{j=1}^{m} C_j.\exp\left(-\frac{Ea_j}{Kb.\theta}\right)} \qquad [10.7]$$

10.4. Summary

– Failure mode analysis and its effects are often major steps in system design.

– This analysis requires knowledge of the failure mode distribution of the components in the system.

– This distribution is often obtained from standard documents such as FMD 2016 (Failure Mode Distribution).

– The distribution of failure modes is independent of time and the life profile to which the system components are subjected.

– This means that an exponential distribution is systematically used; therefore, failure modes resulting from aging mechanisms are not taken into consideration.

– This chapter proposes to define an estimator for the distribution of component failure modes for any type of failure mechanism as well as for systems that undergo maintenance under operating conditions.

11

Feared Events (FTA)

11.1. Introduction

For many industrial applications, some specific events are critical. Examples include the loss of flight controls on an aircraft, radioactive leaks in a nuclear power plant or, even more recently, malfunctioning in a car cruise control. There is obviously no such thing as a "0" risk, and we are constrained to define a maximum probability of occurrence of the event not to be exceeded. The value of this probability will naturally depend on the industrial context but also primarily on the criticality of the event.

In many application cases, its theoretical calculation is made from simplifying assumptions that are rarely explained and particularly not always verified.

When the failure probability is considered, the failure rate should be considered because these two reliability quantities are linked to each other by the following relation:

$$P(t) = 1 - \exp\left(- \int_0^t h(u).du\right) \quad [11.1]$$

This equation is, however, only suitable for maintenance-free systems, which in fact only applies to a few industrial applications.

When the failure rate can be modeled by an exponential distribution, we obtain the classical formula:

$$P(t) = 1 - \exp(-\lambda.t) \quad [11.2]$$

This equation is often used in industrial applications. On the other hand, the failure rate, and therefore the probability, can also depend on physical constraints such as temperature, thermal cycling, humidity, and vibrations. In this case, we can rewrite the previous equation in the following manner, denoting the level of physical stress by X:

$$P(t,X) = 1 - \exp\left(-\int_0^t h(u,X).du\right) \quad [11.3]$$

In this last equation, we consider that the level of physical stress is constant, but very often, during the operational life of the system, this does not occur; therefore, we should instead write it as follows:

$$P(t,X(t)) = 1 - \exp\left(-\int_0^t h(u,X(u)).du\right) \quad [11.4]$$

11.2. Regulatory aspects

The regulatory aspects obviously depend on the industrial context. Consider the example of civil avionics to illustrate our point. These are specified by the European Aviation Safety Agency (EASA) through the document (European Aviation Safety Agency Amendment 2007).

11.2.1. *CS25 framework*

This document presents how the probability of failure is calculated in the Appendix. The calculation of the probability is given by (European Aviation Safety Agency Amendment 2007):

$$\begin{aligned} P_{fight}(\text{Failure}) &= \sum_{j=1}^{n} P_{Phasej}(\text{Failure}) = \sum_{j=1}^{n} P\left(\text{Failure} \mid t \in \left[t_{j-1}, t_j\right]\right) \\ &= 1 - \prod_{i=1}^{n} \exp\left(-\int_{t_{i-1}}^{t_i} \lambda_1(x)\,dx\right) \end{aligned} \quad [11.5]$$

We now explain in detail how the equation was obtained and assume that we are addressing catalectic failures only, that is, failures of accidental origin. In this case, the failure rate does not depend on physical contributions such as temperature or humidity. Equation [11.4] can therefore be written as:

$$P(t) = 1 - \exp\left(-\int_0^t h(u).du\right)$$

Although only catalectic failures are addressed, namely, those that can be modeled by a constant failure rate, few industrial applications exhibit a failure rate that remains the same throughout the operational operation of the product in question. A more realistic example that can be considered is often observed for more or less high numbers of accidental failures depending on the operational activity. This is equivalent to saying that the product failure rates are constant, but only for each operational activity. Indeed, the number of accidental failures can be higher during human activity, product handling, etc. Let us assume that the operational activity of a product can be represented by "n" distinct operational phases. The system fails as soon as a failure occurs during one of the "n" phases. Since these events are independent and mutually exclusive, the following expression can be written:

$$P(t) = P_1(t) \cup P_2(t) \cup \ldots \cup P_{n-1}(t) \cup P_n(t)$$

In this case, the reliability function of the system is written as:

$$R(t) = \prod_{i=1}^{n} R_i(t)$$

If we consider that the "*i*th" phase lasts the time interval $[t_{i-1}, t_i]$ and that $t_0 = 0$, we obtain:

$$R(t) = \prod_{i=1}^{n} exp\left(- \int_{t_{i-1}}^{t_i} \lambda_i . du \right)$$

This is therefore:

$$P(t) = 1 - \prod_{i=1}^{n} exp\left(- \int_{t_{i-1}}^{t_i} \lambda_i . du \right)$$

We indeed return to equation [11.5].

11.2.2. *Problems found*

The previous equation may raise several issues. Clearly, it does not consider aging failure, which results in a time-dependent failure rate. Using equation [11.3] might seem to be an option, but it is only valid for maintenance-free systems since it is based on the failure rate. This creates an issue because most industrial systems are kept in operational conditions. For this equation to be used, we should be certain that there are no failures in the entire pool of systems in operation. This is obviously impossible, namely because of existing catalectic failures that cannot be completely eradicated and which can be observed very early in the operational life of a product.

Consider a situation in which a system periodically performs an operational activity, which can be found in many industrial applications (railways, avionics, etc.). Each element of the failure tree considered can be tested before each operational period. To further illustrate our point, let us assume that we have a single element following an exponential distribution with parameter $\lambda = 0.01$.

When maintenance is possible, the system is tested before each operational activity.

Therefore, when a failure occurs, the system is instantly replaced by a new system. To estimate the probability of occurrence of the feared event during risk time, which is equal to the duration of an operational activity, we use Monte Carlo simulations from within the BlockSim software. This gives us, for the 1,000 simulations planned, the average number of failures per unit of time, which is every 10 h.

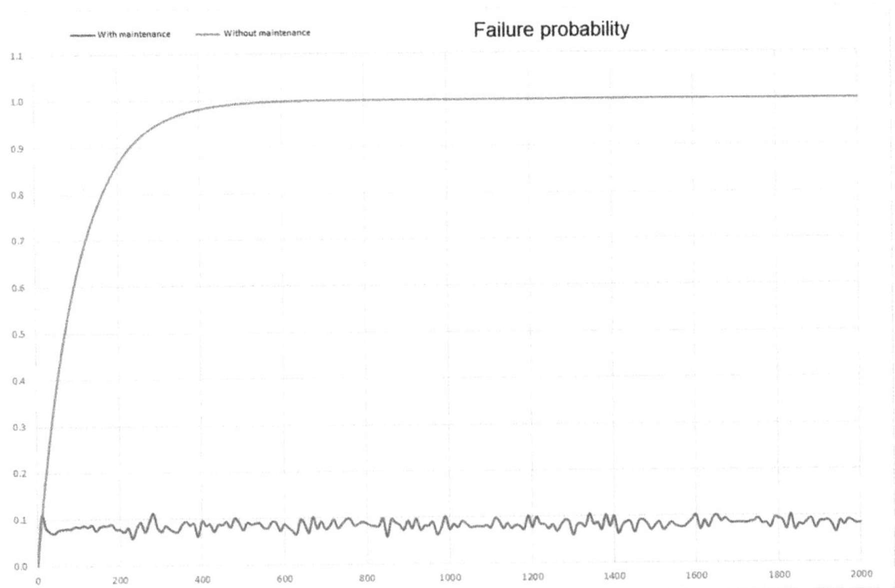

Figure 11.1. *Probability of exponential distribution failure $\lambda = 0.01$ with and without maintenance. For a color version of this figure, see www.iste.co.uk/bayle/montecarlo.zip*

Figure 11.1 shows the probabilities of failure of this element with and without maintenance.

The two probabilities clearly differ, which means that maintenance has a positive impact, as expected. We want to calculate the probability that the system will fail during its operational activity, knowing that it was functional when it started. To this end, consider the random variable X defined by:

$$X = \mathrm{N}(t + \Delta t) - N(t)$$

where:

– Δt represents the risk time;

– N(t) is the average number of failures of the element under consideration.

If it is assumed that the risk time Δt is small enough so that the system does not fail more than once in the interval [t; t + Δt], then we can consider that the variable X follows a Bernoulli law since it can only take a value of 0 or 1. Consequently, the probability of failure $P_{\Delta t}$ of X at time "t" can be written by linearity of the mathematical expectation E (Bayle 2019) as:

$$P_{\Delta t}(t) = \mathrm{E}[\mathrm{N}(t + \Delta t) - N(t)] = \mathrm{E}[\mathrm{N}(t + \Delta t)] - \mathrm{E}[N(t)]$$

On the other hand, we observed in Chapter 3 (on system reliability) that:

$$\mathrm{E}[\mathrm{N}(t)] = \int_0^t Rocof(u).du$$

from which:

$$P_{\Delta t}(t) = \int_0^{t+\Delta t} Rocof(u).du - \int_0^t Rocof(u).du$$

namely:

$$P_{\Delta t}(t) = \int_0^{\Delta t} Rocof(u).du + \int_0^t Rocof(u).du - \int_0^{t+} Rocof(u).du$$

That is, finally:

$$P_{\Delta t}(t) = \int_t^{t+\Delta t} Rocof(u).du \qquad [11.6]$$

We then compute the limit as t approaches $+\infty$ of this probability. We have:

$$lim_{t\to+\infty}P_{\Delta t}(t) = lim_{t\to+\infty}\mathrm{E}[\mathrm{N}(t+\Delta t)] - \mathrm{E}[N(t)]$$

that is:

$$lim_{t\to+\infty}P_{\Delta t}(t) = lim_{t\to+\infty}\mathrm{E}[\mathrm{N}(t+\Delta t)] - lim_{t\to+\infty}\mathrm{E}[N(t)]$$

Now, we know that (Basu and Rigdon 2000): $lim_{t\to+\infty}N(t) = \frac{t}{MTTF}$; hence:

$$P_{\Delta t}(\infty) = \frac{t+\Delta t}{MTTF} - \frac{t}{MTTF}$$

and, finally:

$$\boldsymbol{P_{\Delta t}(\infty) = \frac{\Delta t}{MTTF}} \qquad [11.7]$$

In the following sections, we will study the expression of this conditional probability for different models of system reliability from a theoretical point of view. On the other hand, we compare these results with those obtained by Monte Carlo simulations. This method leads to obtaining the average number of failures for each time step. We then need to subtract only the average number of failures between two consecutive time steps to obtain an estimate of the probability of conditional failure.

11.2.3. *Using an exponential distribution*

Let us assume that system reliability can be modeled by an exponential distribution with parameter λ. We already saw that this model is particularly well suited to catalectic failures. We also know that in this case, the failure occurrence rate is given by (Bayle 2019):

$$Rocof(t) = \lambda$$

Equation [11.6] is then written as:

$$P_{\Delta t}(t) = \int_t^{t+\Delta t} \lambda.du = \lambda.\Delta t \qquad [11.8]$$

Therefore, the probability that the system will fail during the next operational activity, knowing that it was functional when it started, depends solely on the parameter λ of the exponential distribution and the duration of the operational phase Δt.

REMARK.– If we compare equation [11.8] with the formulation of the maintenance-free systems of equation [11.3], they appear to be similar.

However, there is a fundamental difference: one is time dependent, whereas the other is simply dependent on the risk time Δt.

11.2.4. *Using a normal distribution*

Let us assume that system reliability can be modeled by a normal distribution with parameters μ and σ. We already saw that this model would be suitable for age-related failures under certain conditions (Bayle 2019).

We also know that in this case, the Rocof is given by:

$$Rocof(t) = \sum_{i=1}^{+\infty} \varphi(t, i.\mu, \sqrt{i}.\sigma)$$

As a result, we obtain:

$$P_{\Delta t}(t) = \int_{t}^{t+\Delta t} \sum_{i=1}^{+\infty} \varphi(u, i.\mu, \sqrt{i}.\sigma).du$$

that is:

$$P_{\Delta t}(t) = \sum_{i=1}^{+\infty} \int_{t}^{t+\Delta t} \varphi(u, i.\mu, \sqrt{i}.\sigma).du$$

Finally, where ϕ represents the probability of failure of the normal distribution:

$$P_{\Delta t}(t) = \sum_{i=1}^{+\infty}[\phi(t + \Delta t, i.\mu, \sqrt{i}.\sigma) - \phi(t, i.\mu, \sqrt{i}.\sigma)] \quad [11.9]$$

The asymptotic value is obtained from equation [11.7], namely:

$$P_{\Delta t}(\infty) = \frac{\Delta t}{\mu} \quad [11.10]$$

EXAMPLE 11.1.–

Normal distribution with parameters μ = 100; σ = 20; Δt = 20. We obtain the following graph (Figure 11.2).

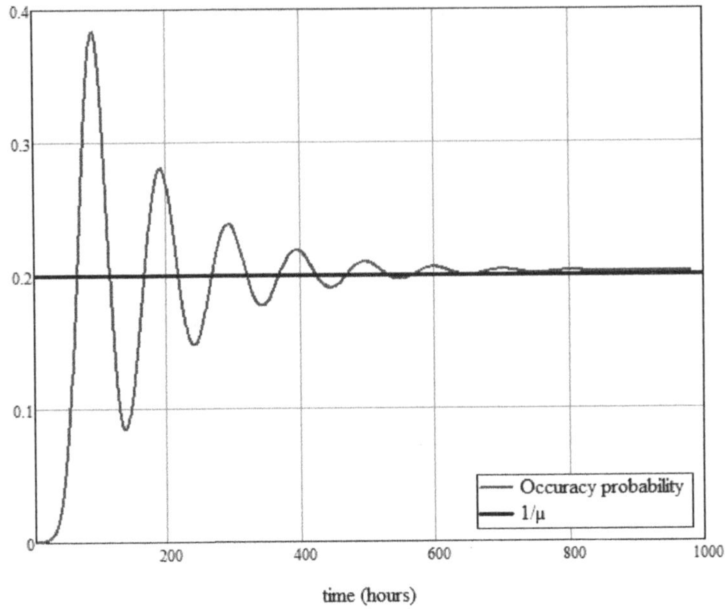

Figure 11.2. *Occurrence probability for a normal distribution $\mu = 100$; $\sigma = 20$; $\Delta t = 20$. For a color version of this figure, see www.iste.co.uk/bayle/montecarlo.zip*

REMARK.– We see that this conditional probability has an asymptotic value of 0.2 for a Δt of 20 h. The conditional probability will reach its smallest values for shorter periods of the operational phase Δt.

This result also makes sense, since the shorter Δt is, the less likely the system is to fail.

11.3. Probability of the occurrence of a feared event

For industrial systems, which are generally complex, a feared event is defined by a "fault tree" and not a single element, as was the case in the previous section, which presented the theoretical approach.

We therefore present the calculation of this conditional probability for simple events such as AND and OR gates.

11.3.1. *OR gate with element testing*

Only one element must be faulty for the feared event to be observed. If we consider "n" independent and mutually exclusive elements, the probability of observing the feared event is given by:

$$P_{\Delta t}(t) = \sum_{i=1}^{n} P_i(t, \Delta t)$$

that is:

$$P_{\Delta t}(t) = \sum_{i=1}^{n} \int_t^{t+\Delta t} Rocof_i(u).du \qquad [11.11]$$

The asymptotic value is equal to:

$$P_{\Delta t}(\infty) = lim_{t \to +\infty} \sum_{i=1}^{n} \int_t^{t+\Delta t} Rocof_i(u).du$$

that is:

$$P_{\Delta t}(\infty) = \sum_{i=1}^{n} lim_{t \to +\infty} \int_t^{t+\Delta t} Rocof_i(u).du$$

that is:

$$P_{\Delta t}(\infty) = \sum_{i=1}^{n} lim_{t \to +\infty} \left(\int_0^{t+\Delta t} Rocof_i(u).du - \int_0^{t} Rocof_i(u).du \right)$$

that is:

$$P_{\Delta t}(\infty) = \sum_{i=1}^{n} \left(lim_{t \to +\infty} \left(\int_0^{t+\Delta t} Rocof_i(u).du \right) - lim_{t \to +\infty} \left(\int_0^{t} Rocof_i(u).du \right) \right)$$

namely:

$$P_{\Delta t}(\infty) = \sum_{i=1}^{n} \left(\frac{t + \Delta t}{MTTF_i} - \frac{t}{MTTF_i} \right)$$

and finally:

$$P_{\Delta t}(\infty) = \sum_{i=1}^{n} \frac{\Delta t}{MTTF_i} \quad [11.12]$$

Exponential distributions

It is known that $Rocof_i(t) = \lambda_i$, from which:

$$P_{\Delta t}(t) = \Delta t. \sum_{i=1}^{n} \lambda_i = P_{\Delta t}(\infty) \quad [11.13]$$

Normal distributions

It is known that $Rocof_i(t) = \sum_{j=1}^{+\infty} \varphi(t, j.\mu_i, \sqrt{j}.\sigma_i)$; hence:

$$P_{\Delta t}(t) = \sum_{i=1}^{n} \int_{t}^{t+\Delta t} \sum_{j=1}^{+\infty} \varphi(u, j.\mu_i, \sqrt{j}.\sigma_i).du$$

which is finally:

$$P_{\Delta t}(t) = \sum_{i=1}^{n} \sum_{j=1}^{+\infty} \left(\phi(t+\Delta t, j.\mu_i, \sqrt{j}.\sigma_i) - \phi(t, j.\mu_i, \sqrt{j}.\sigma_i)\right) \quad [11.14]$$

The asymptotic value is given by:

$$P_{\Delta t}(\infty) = \Delta t. \sum_{i=1}^{n} \frac{1}{\mu_i} \quad [11.15]$$

11.3.2. *OR gate without element testing*

The elements are not tested, but since only one of them has to fail to observe the feared event, testing them is useless. On the other hand, when an element fails, and therefore when the feared event is observed, a decision can be made either to renew the defective element only or to renew them all.

In the first case, we are faced again with the case previously analyzed. For the other possibility, let us calculate the reliability function of the feared event.

We have:

$$R_E(t) = \prod_{i=1}^{n} R_i(t)$$

The probability density is therefore written as:

$$f_E(t) = \sum_{i=1}^{n} \left[f_i(t). \prod_{\substack{j=1 \\ j \neq i}}^{n} R_j(t) \right]$$

Exponential distributions

$$f_E(t) = \sum_{i=1}^{n} \left[\lambda_i. exp(-\lambda_i. t) \prod_{\substack{j=1 \\ j \neq i}}^{n} exp(-\lambda_j. t) \right]$$

that is, finally:

$$f_E(t) = \sum_{i=1}^{n} \lambda_i . exp\left(- \sum_{i=1}^{n} \lambda_i . t \right)$$

The Rocof is therefore equal to:

$$Rocof(t) = \sum_{i=1}^{n} \lambda_i$$

Consequently:

$$P_{\Delta t}(t) = \Delta t. \sum_{i=1}^{n} \lambda_i = P_{\Delta t}(\infty)$$

11.3.3. *AND gate with element testing*

To observe a feared event, all of the elements must be defective. Under the assumption of independence of the “n” elements, we can write the following:

$$P_{\Delta t}(t) = \prod_{i=1}^{n} P_{\Delta t_i}(t)$$

This is done according to equation [11.6]:

$$P_{\Delta t}(t) = \prod_{i=1}^{n} \left[\int_t^{t+\Delta t} Rocof_i(u). du \right] \qquad [11.16]$$

If the elements are identical, we obtain:

$$P_{\Delta t}(t) = \left[\int_{t}^{t+\Delta t} Rocof(u).du\right]^{n}$$

The asymptotic value is given by:

$$P_{\Delta t}(\infty) = lim_{t\to+\infty} \prod_{i=1}^{n} \left[\int_{t}^{t+\Delta t} Rocof_i(u).du\right]$$

that is:

$$P_{\Delta t}(\infty) = \prod_{i=1}^{n} \left[lim_{t\to+\infty} \int_{t}^{t+\Delta t} Rocof_i(u).du\right]$$

that is:

$$P_{\Delta t}(\infty) = \prod_{i=1}^{n} \left[\frac{\Delta t}{MTTF_i}\right]$$

that is:

$$P_{\Delta t}(\infty) = \Delta t^{n} \prod_{i=1}^{n} \left[\frac{1}{MTTF_i}\right] \quad [11.17]$$

If all of the elements are the same, we obtain:

$$P_{\Delta t}(\infty) = \left(\frac{\Delta t}{MTTF}\right)^{n}$$

Exponential distributions

We can write:

$$P_{\Delta t}(t) = \prod_{i=1}^{n} \left[\int_{t}^{t+\Delta t} \lambda_i.du\right]$$

namely:

$$P_{\Delta t}(t) = \Delta t^n \prod_{i=1}^{n} \lambda_i$$

If all of the elements are the same, we obtain:

$$P_{\Delta t}(t) = (\lambda . \Delta t)^n$$

Normal distributions

We can write:

$$P_{\Delta t}(t) = \prod_{i=1}^{n} \int_{t}^{t+\Delta t} \sum_{j=1}^{+\infty} \varphi\left(u, j.\mu_i, \sqrt{j}.\sigma_i\right).du$$

If all of the elements are the same, we obtain:

$$P_{\Delta t}(t) = \left[\int_{t}^{t+\Delta t} \sum_{j=1}^{+\infty} \varphi\left(u, j.\mu_i, \sqrt{j}.\sigma_i\right).du\right]^n$$

The asymptotic value is given by:

$$P_{\Delta t}(\infty) = lim_{t\to+\infty}\left[\int_{t}^{t+\Delta t} \sum_{j=1}^{+\infty} \varphi\left(u, j.\mu_i, \sqrt{j}.\sigma_i\right).du\right]^n$$

that is:

$$P_{\Delta t}(\infty) = \left[lim_{t\to+\infty} \int_{t}^{t+\Delta t} \sum_{j=1}^{+\infty} \varphi\left(u, j.\mu_i, \sqrt{j}.\sigma_i\right).du\right]^n$$

that is:

$$P_{\Delta t}(\infty) = \left[lim_{t\to+\infty} \sum_{j=1}^{+\infty} \phi\left(t + \Delta t, j.\mu_i, \sqrt{j}.\sigma_i\right) - \sum_{j=1}^{+\infty} \phi\left(t, j.\mu_i, \sqrt{j}.\sigma_i\right)\right]^n$$

that is:

$$P_{\Delta t}(\infty) = \frac{\Delta t^n}{\prod_{i=1}^{n} \mu_i}$$

11.3.4. *AND gate without element testing*

A feared event is observed when all of the elements are defective. However, since the elements are not tested, this observation can be made only if all of the elements are defective. Let us calculate the probability of occurrence of the feared event, namely:

$$F_E(t) = \prod_{i=1}^{n} F_i(t)$$

The corresponding probability density is therefore:

$$f_E(t) = \frac{\partial}{\partial t} \prod_{i=1}^{n} F_i(t)$$

and still:

$$f_E(t) = \sum_{i=1}^{n} \left[f_i(t). \prod_{\substack{j=1 \\ j \neq i}}^{n} F_j(t) \right]$$

Exponential distributions

We can write:

$$f_E(t) = \sum_{i=1}^{n} \left[\lambda_i . exp(-\lambda_i . t). \prod_{\substack{j=1 \\ j \neq i}}^{n} [1 - exp(-\lambda_j . t)] \right]$$

Even for exponential laws, there are no simple expressions of the Rocof. However, if the elements are identical, we obtain:

$$f_E(t) = \sum_{i=1}^{n} \left[\lambda . exp(-\lambda . t). \prod_{\substack{j=1 \\ j \neq i}}^{n} [1 - exp(-\lambda . t)] \right]$$

that is:

$$f_E(t) = \text{n}.\lambda.exp(-\lambda.t).\left(1 - exp(-\lambda.t)\right)^{n-1}$$

There are also no simple expressions for the Rocof and therefore for the probability of occurrence of the feared event.

Let us calculate the MTTF:

$$MTTF = \int_0^{+\infty} \left(1 - \prod_{i=1}^{n}\left(1 - R_i(t)\right)\right).dt$$

The analytical calculation of this integral is not known, and the asymptotic value has no explicit solutions.

11.4. Practical application

The objective is to estimate the average probability of failure of a system with a double failure. For example, a system can include a function and its monitoring.

The function can be the generation of a supply voltage and monitoring, namely, it is able to detect if this voltage exceeds a certain high threshold, allowing the function to be cut off and/or to send a status signal.

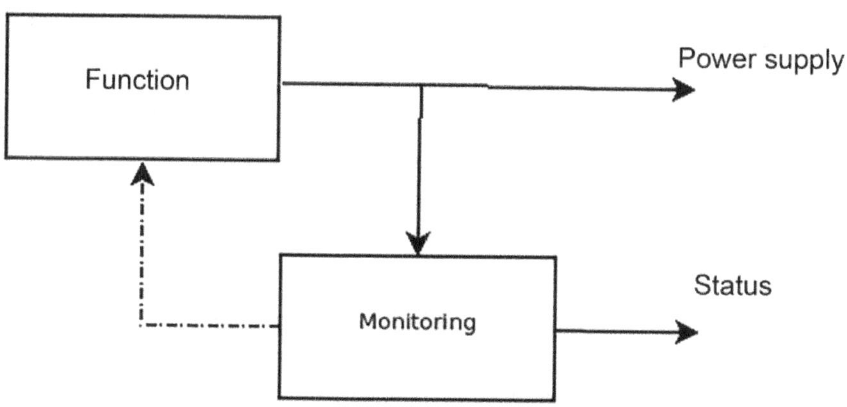

Figure 11.3. *Illustration of a double failure system*

We want to determine the probability that the function and its monitoring are simultaneously faulty with the following assumption:

– monitoring must be faulty before the function;

– the proper functioning of the monitoring system is tested over a period of time longer than the inactivity of the monitored function. This is called a latent fault.

This is equivalent to an AND gate in a fault tree, as shown in Figure 11.4.

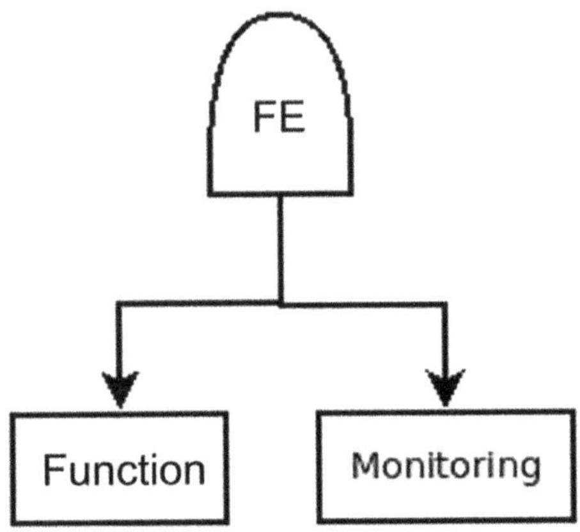

Figure 11.4. *Illustration of the corresponding feared event*

The principle behind a simulation is as follows:

– generate a monitoring failure time (Tmon);

– generate function (Tfon) failure times by time accumulation up to a determined final time;

– keep only Tfon greater than Tmon;

– determine the frequency table of these times by classes of width FH.

As an illustration, for a first simulation, if Tmon = 450 and Tfon = {50, 340, 620, 810}, then the retained times will be {620, 810} because they are greater than 450, and the frequency table FH=100 will be as follows:

FH	100	200	300	400	500	600	700	800	900
Tfon	0	0	0	0	0	0	1	0	1

For a second simulation, if Tmon = 245 and Tfon = {100, 430, 680}, then the retained times will be {430, 680}, and the updated frequency table will be as follows:

FH	100	200	300	400	500	600	700	800	900
Tfon	0	0	0	0	1	0	2	0	1

The unit probability is obtained by dividing each class by the number of simulations performed (two in this case):

FH	100	200	300	400	500	600	700	800	900
P (/FH)	0	0	0	0	0.5	0	1	0	0.5

Finally, it is possible to normalize each class by dividing them by FH = 100, which gives a probability to the flight hour:

FH	100	200	300	400	500	600	700	800	900
P (/h)	0	0	0	0	0.005	0	0.01	0	0.005

By iterating this process several thousand or even tens of thousands of times, the probabilities are refined.

The following timetable gives the general principle of the simulations:

If we take the parameters MTBF function = 1,000 h, MTBF_Monitoring = 10,000 h and risk time Tvol = 10 h, we obtain the simulation result in Figure 11.6.

REMARK.– The curve is strongly noisy because there are not enough simulated failures; therefore, the number of simulations is not large enough. In this example, we see some limitations in Monte Carlo simulations in safety because, in practice, the MTBF values can be an order of magnitude larger than those considered in this example. This means that an even greater number of simulations will be needed, and consequently, simulation times that may become unfeasible.

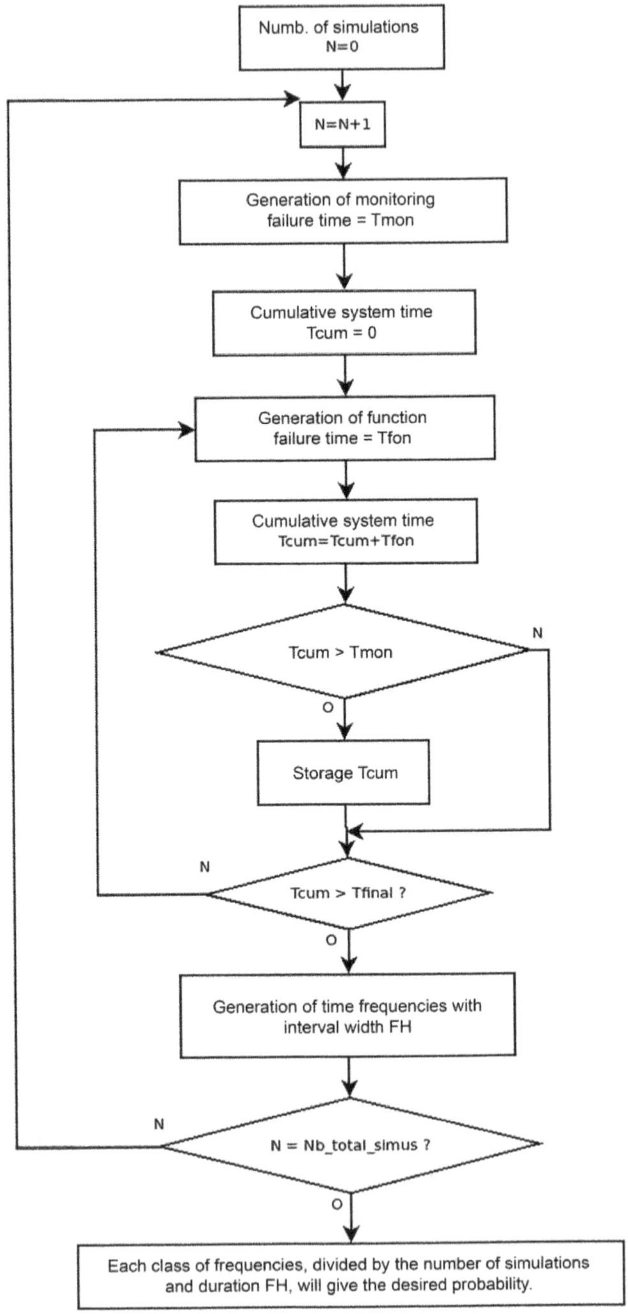

Figure 11.5. *Simulation flow chart of the feared event*

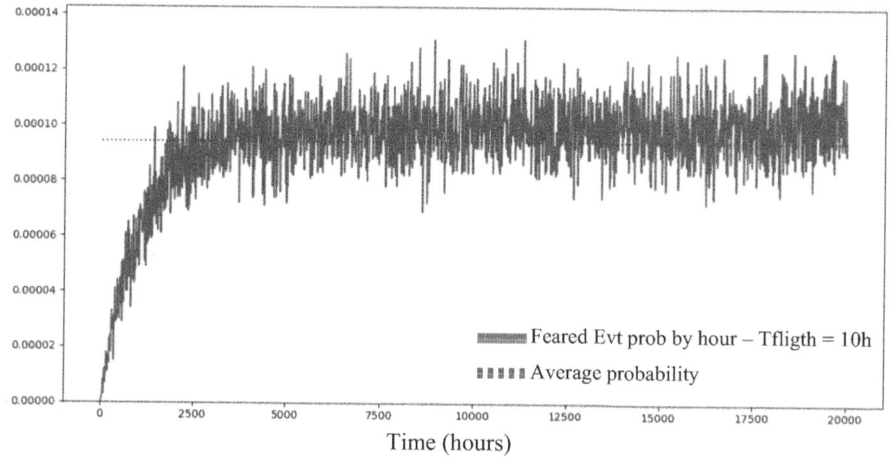

Figure 11.6. *Simulation of the probability of the feared event. For a color version of this figure, see www.iste.co.uk/bayle/montecarlo.zip*

11.5. Summary

– Safety is an analysis that consists of estimating the probability of occurrence of critical feared events.

– This analysis is usually performed with the assumption that the failure rate is constant, as seen in the regulatory aspects of the CS25 in the avionics domain.

– The effect of maintenance is not taken into account, even though it is present in most industrial applications.

– In addition, the equations defining the different probabilities included in a fault tree are based on the fact that the state of the system is known before each operational phase, which sets the underlying assumption that a test is systematically carried out and that any failed system elements are then replaced by new ones.

– This chapter proposes a theoretical principle to cover situations where component aging mechanisms exist and where ideal maintenance actions (instantaneous replacement of failed elements with new ones) are performed.

– This principle will then be developed for AND and OR gates.

– We then present a practical case study of a system with a main function and monitoring.

– Nonetheless, in practice, for very high levels of reliability, Monte Carlo simulations can lead to excessively long simulation times and are not really interesting in practice.

Appendices

A.1. Appendix 1: calculation of the MTTF of a k/n redundancy for an exponential distribution

The reliability function of an active k/n redundancy is given by Rausand and Hoyland (2004):

$$R_{kn}(t) = \sum_{i=k}^{n} C_n^i . exp(-\lambda . i . t) . \left(1 - exp(-\lambda . t)\right)^{n-i}$$

The corresponding MTTF is therefore given by:

$$MTTF_{kn} = \int_0^{+\infty} R_{kn}(t) . dt = \int_0^{+\infty} \sum_{i=k}^{n} C_n^i . exp(-\lambda . i . t) . \left(1 - exp(-\lambda . t)\right)^{n-i} . dt$$

that is:

$$MTTF_{kn} = \sum_{i=k}^{n} C_n^i . \int_0^{+\infty} exp(-\lambda . i . t) . \left(1 - exp(-\lambda . t)\right)^{n-i} . dt$$

Let $u = exp(-\lambda . t)$ and therefore $u = exp(-\lambda . t)$ from which:

$$MTTF_{kn} = \sum_{i=k}^{n} C_n^i . \int_1^{0} u^i . (1 - u)^{n-i} . \frac{du}{-\lambda . u}$$

That is, since $MTTF_e = \frac{1}{\lambda}$:

$$MTTF_{kn} = MTTF_e.\sum_{i=k}^{n} C_n^i.\int_0^1 u^{i-1}.(1-u)^{n-i}.du$$

It is known that the beta function is defined by:

$$B(x,y) = \int_0^1 u^{x-1}.(1-u)^{y-1}.du = \frac{\Gamma(x).\Gamma(y)}{\Gamma(x+y)}$$

Consequently, the MTTF can be written as:

$$MTTF_{kn} = MTTF_e.\sum_{i=k}^{n} C_n^i.\frac{\Gamma(i).\Gamma(n-i+1)}{\Gamma(i+n-i)}$$

That is, since $\Gamma(n) = (n-1)!$, given that n is a natural number:

$$MTTF_{kn} = MTTF_e.\sum_{i=k}^{n} \frac{n!}{i!\,(n-i)!}.\frac{(i-1)!\,(n-i)!}{n!}$$

and, finally:

$$MTTF_{kn} = MTTF_e.\sum_{i=k}^{n} \frac{1}{i}$$

To achieve this result, we used the following expression:

$$\sum_{i=1}^{p} \frac{C_p^i.(-1)^{i+1}}{i} = \sum_{i=1}^{p} \frac{1}{i}$$

This can be demonstrated using recurrence. Indeed, for p = 1, we have:

$\sum_{i=1}^{p} \frac{C_p^i.(-1)^{i+1}}{i} = \frac{C_1^1.(-1)^{1+1}}{1} = 1$ but also $\sum_{i=1}^{p} \frac{1}{i} = \frac{1}{1} = 1$.

The equality is therefore true at rank 1.

Suppose that the equality is true at rank "p". At rank "p+1", we have:

$$\sum_{i=1}^{p+1} \frac{C_{p+1}^{i}.(-1)^{i+1}}{i} = \sum_{i=1}^{p} \frac{C_{p}^{i}.(-1)^{i+1}}{i} + \frac{C_{p+1}^{p+1}.(-1)^{p+1+1}}{p+1}$$

That is:

$$\sum_{i=1}^{p+1} \frac{C_{p+1}^{i}.(-1)^{i+1}}{i} = \sum_{i=1}^{p} \frac{1}{i} + \frac{1}{p+1}$$

and, finally:

$$\sum_{i=1}^{p+1} \frac{C_{p+1}^{i}.(-1)^{i+1}}{i} = \sum_{i=1}^{p+1} \frac{1}{i}$$

A.2. Appendix 2: calculation of the asymptotic value of the Rocof in a serial system with perfect maintenance

Each serial system element is a renewal process. We thus have a superposition of "s" renewal process and can therefore write that:

$$\mathrm{E}[N_S(t)] = \mathrm{E}\left[\sum_{i=1}^{s} N_i(t)\right]$$

and therefore, by linearity of the mathematical expectation:

$$\frac{\partial}{\partial t}\mathrm{E}[N_S(t)] = \sum_{i=1}^{s} \frac{\partial}{\partial t} E[N_i(t)]$$

That is furthermore:

$$Rocof_S(t) = \sum_{i=1}^{s} Rocof_i(t)$$

and thereby:

$$Rocof_\infty = lim_{t\to+\infty} \sum_{i=1}^{s} Rocof_i(t)$$

That is:

$$Rocof_{\infty} = \sum_{i=1}^{s} lim_{t\to+\infty} Rocof_i(t)$$

and finally:

$$Rocof_{\infty} = \sum_{i=1}^{s} \frac{1}{MTTF_i}$$

REMARK.– When all of the serial elements follow an exponential law, we get:

$$Rocof_{\infty} = \sum_{i=1}^{s} \lambda_i$$

Again, we find the expression of the failure rate of a serial system because the superposition of "s" renewal process is a renewal process when the underlying probability density is an exponential distribution.

A.3. Appendix 3: calculation of the probability density of a serial system containing "s" elements

Consider a parallel system of "p" elements with probability density f. The probability density of the system f_p is then given by:

$$f_s(t) = \sum_{i=1}^{s} \left(f_i(t). \prod_{\substack{j=1 \\ j\neq i}}^{s} R_j(t) \right)$$

The demonstration is done using recurrence. Let s = 1 and therefore: $f_s(t) = f(t)$. The law is therefore true in rank 1. Suppose it is true at rank "s". We thus have:

$$f_s(t) = \sum_{i=1}^{s} f_i(t). \prod_{\substack{j=1 \\ j\neq i}}^{s} R_j(t)$$

For "s + 1" elements, we will have:

$$R(t) = R_s(t).R_{s+1}(t)$$

from which:

$$f(t) = -\frac{d}{dt}R(t) = -\frac{d}{dt}\left(R_s(t).R_{s+1}(t)\right) = f_s(t).R_{s+1}(t) + R_s(t).f_{s+1}(t)$$

That is:

$f(t) = \left[\sum_{i=1}^{s}\left(f_i(t).\prod_{\substack{j=1\\j\neq i}}^{s}R_j(t)\right)\right].R_{s+1}(t) + f_{s+1}(t).\prod_{j=1}^{s}R_j(t)$ because $R_s(t) = \prod_{j=1}^{s}R_j(t)$ and furthermore

$$f(t) = \sum_{i=1}^{s}\left(f_i(t).\prod_{\substack{j=1\\j\neq i}}^{s}R_j(t)\right).R_{s+1}(t) + f_{s+1}(t).\prod_{j=1}^{s}R_j(t)$$

that is finally:

$$f_{s+1}(t) = \sum_{i=1}^{s+1}\left(f_i(t).\prod_{\substack{j=1\\j\neq i}}^{s+1}R_j(t)\right)$$

hence the result.

A.4. Appendix 4: sum of exponential distributions

It is known (Basu and Rigdon 2000) that the sum of n random variables was defined by the convolution product of its probability density "n" times by itself. To carry out this calculation, the integral definition of the convolution product could be employed but the computation is quite long and tedious. It is preferable to use the following property of the Laplace transform:

$$\mathcal{L}\left(\mathrm{f}(t) * g(t)\right) = \mathcal{L}\left(f(t)\right).\mathcal{L}\left(g(t)\right)$$

In our case, this leads to:

$$\mathcal{L}\left(f^{<n>}(t)\right) = \left(\mathcal{L}\left(f(t)\right)\right)^n$$

The density of the exponential distribution with parameter λ is given by:

$$f(t) = \lambda.\exp(-\lambda.t).\beth_{t\geq 0}$$

from which:

$$\mathcal{L}\left(fe^{<n>}(t)\right) = (\lambda.\exp(-\lambda.t).\beth_{t\geq 0})^n$$

Moreover, $\mathcal{L}(\lambda.\exp(-\lambda.t).\beth_{t\geq 0}) = \frac{1}{p+\lambda}$, where "p" is the Laplace operator. We therefore obtain:

$$\mathcal{L}\left(fe^{<n>}(t)\right) = \left(\frac{\lambda}{p+\lambda}\right)^n$$

The inverse Laplace transform of $\left(\frac{\lambda}{p+\lambda}\right)^n$ is given by:

$\mathcal{L}^{-1}\left(\left(\frac{1}{p+\lambda}\right)^n\right) = \frac{t^{n-1}}{\Gamma(n)}.\exp(-\lambda.t)$ where Γ is the Gamma function, from which:

$$fe^{<n>}(t) = \frac{\lambda^n.t^{n-1}.\exp(-\lambda.t)}{\Gamma(n)}$$

We then recognize the expression of the probability density of a gamma distribution of parameters n and λ.

A.5. Appendix 5: failure rate for a serial system

The failure rate h(t) is related to the survival function R(t) by the following equation:

$$R(t) = \exp\left(-\int_0^t h(u).du\right)$$

For a serial system with "s" independent elements, it is known that:

$$R(t) = \prod_{i=1}^{s} R_i(t)$$

These two equations enable us to write that:

$$R(t) = \prod_{i=1}^{s} \exp\left(- \int_0^t h_i(u).du \right)$$

that is:

$$R(t) = \exp\left(- \sum_{i=1}^{s} \int_0^t h_i(u).du \right)$$

and, finally:

$$R(t) = \exp\left(- \int_0^t \sum_{i=1}^{s} h_i(u) \right)$$

Denoting h_s the serial system failure rate, we obtain:

$$h_s(t) = \sum_{i=1}^{s} h_i(u)$$

A.6. Appendix 6: relation between failure rate and reliability function

By definition of the default rate, we know that:

$$h(t) = \frac{f(t)}{\mathrm{R}(t)}$$

On the other hand, we have:

$$f(t) = \frac{\mathrm{d}}{\mathrm{dt}} F(t) = \frac{\mathrm{d}}{\mathrm{dt}} \big(1 - R(t)\big) = -\frac{\mathrm{d}}{\mathrm{dt}} R(t)$$

from which:

$$h(t) = \frac{-\frac{\mathrm{d}}{\mathrm{dt}}R(t)}{\mathrm{R}(t)}$$

Yet it is known that: $\frac{d}{dx}\ln(u(x)) = \frac{\frac{d}{dx}u(x)}{\mathrm{u}(x)}$ therefore:

$$h(t) = -\frac{d}{dt}ln(R(t))$$

Computing the integration of this last equation, we can write:

$$\int_0^t h(u).du = -ln(R(t)) + C$$

that is:

$$R(t) = exp\left(C - \int_0^t h(u).du\right)$$

that is:

$$R(t) = \mathrm{K}.exp\left(-\int_0^t h(u).du\right)$$

Since the reliability function R(t) is the probability of being functional at time "t", we can write that:

$$R(t = 0) = 1$$

Since all components are functional at time t = 0 from which K = 1, and finally:

$$R(t) = exp\left(-\int_0^t h(u).du\right)$$

References

Akaike, H. (1974). A new look at the statistical model identification. *IEEE Transactions on Automatic Control*, 19(6), 716–723.

Arrhenius, S. (1889). Uber die Reaktionsgeschwindigkeit bei der Inversion von Rohrzucker durch Säuren. *Zeitschrift für Physikalische Chemie*, 4U(1), 226.

Basu, A.P. and Rigdon, S.E. (2000). *Statistical Method for the Reliability of Repairable System*. John Wiley & Sons, New York.

Bayle, F. (2019). *Reliability of Maintained Systems Subjected to Wear Failure Mechanisms*. ISTE Ltd, London and John Wiley & Sons, New York.

Bayle, F. (2021). *Product Maturity 2*. ISTE Ltd, London and John Wiley & Sons, New York.

Bondyopadhyay, P.K. (1998). W=Shockley, the transistor pioneer-portrait of an inventive genius. *Proceedings of the IEEE*, 86(1), 191–217. doi: 10.1109/5.658771.

Crow, L.H. (1974). Reliability analysis for complex, repairable systems. Technical Report 138, AMSAA, Aberdeen, MD.

European Aviation Safety Agency Amendment (2007). Certification specifications for large aeroplanes. Document, European Aviation Safety Agency Amendment, Cologne.

Gaudoin, O. (2007). *Modélisation aléatoire en fiabilité des logiciels*. Hermes Science-Lavoisier, Paris.

IMDR (2022). FIDES Guide Edition A – Reliability methodology for electronic systems. Institut pour la Maitrise des Risques (IMDR), Paris.

Lall, P., Pecht, M., Hakim, E. (1997). *Influence of Temperature on Microelectronics and System Reliability*. CRC Press, Boca Raton.

Nichicon Corporation (2000). General descriptions of aluminium electrolytic capacitors. Technical Notes, Nichicon Corporation, Kyoto.

Nikulin, M., Greville-Réache, L., Couallier, V. (2007). *Statistiques des essais accélérés*. Hermes Science-Lavoisier, Paris.

Pecht, M. and Das, D. (2020). Evaluation and comparison of FIDES and PoF-Based EEE part reliability assessment. Report, University of Maryland, College Park, MD.

Quanterion Solutions Incorporated (2016). *Failure Mode Distribution*. Quanterion Solutions Incorporated, New York.

Rausand, M. and Hoyland, A. (2004). *System Reliability Theory: Models, Statistical Methods and Applications*. John Wiley & Sons, New York.

Sedyakin, N.M. (1966). About the one physical principle in theory of reliability. *Proceedings of Academy of Sciences of URSS*, 3, 80–87.

Texas Instruments (2021). FIT Rate, failure mode distribution LM61440-Q1 [Online] Available at: www.ti.com/lit/fs/snva953/snva953.pdf?ts=1754401638141.

Villemeur, A. (1988). *Sûreté de Fonctionnement des systèmes industriels*. Éditions Eyrolles, Paris.

Index

G, K

L, M

O, P

Q, R

S, T, U

Other titles from

iSTE

in

Mechanical Engineering and Solid Mechanics

2025

LEROY Maurice

Rheology, Physical and Mechanical Behavior of Materials 4: Rigidity and Resistance of Composite Materials, Sizings of Laminate

Rheology, Physical and Mechanical Behavior of Materials 5: Fractures and Fatigue of Metal and Composites

EL HAMI Abdelkhalak

Methods and Applications of Artificial Intelligence: Dynamic Response, Learning, Random Forest, Linear Regression, Interoperability, Additive Manufacturing and Mechatronics

(Artificial Intelligence in Mechanics Set – Volume 2)

EL HAMI Abdelkhalak, RADI Bouchaïb

Multi-physics Optimization: Mechanics, Fluid Interaction Structure, Shaping, Stochastic Finite Elements, Random Vibrations, Fatigue

(Reliability of Multiphysical Systems Set – Volume 20)

2024

LE VAN Anh

Contact in Structural Mechanics: A Weighted Residual Approach

LEROY Maurice

Rheology, Physical and Mechanical Behavior of Materials 2: Working at High Strain Rates, Forming, Sintering, Joining and Welding

Rheology, Physical and Mechanical Behavior of Materials 3: Rigidity and Resistance of Materials, Sizings, Pieces and Structures

SAOUAB Abdelghani, CHAMPMARTIN Stéphane, KHALID NACIRI Jaafar

Some Complex Phenomena in Fluid and Solid Mechanics

(Mathematical and Mechanical Engineering Set – Volume 13)

2023

EL HAMI Abdelkhalak

Uncertainty and Artificial Intelligence: Additive Manufacturing, Vibratory Control, Agro-composite, Mechatronics

(Artificial Intelligence in Mechanics Set – Volume 1)

EL HAMI Abdelkhalak, EID Mohamed

Reliability-based Modeling of System Performance

(Reliability of Multiphysical Systems Set – Volume 19)

EL HAMI Abdelkhalak, DELAUX David, GRZESKOWIAK Henri

Applied Reliability for Industry 1: Predictive Reliability for the Automobile, Aeronautics, Defense, Medical, Marine and Space Industries

(Reliability of Multiphysical Systems Set – Volume 16)

Applied Reliability for Industry 2: Experimental Reliability for the Automobile, Aeronautics, Defense, Medical, Marine and Space Industries

(Reliability of Multiphysical Systems Set – Volume 17)

Applied Reliability for Industry 3: Operational Reliability for the Automobile, Aeronautics, Defense, Medical, Marine and Space Industries

(Reliability of Multiphysical Systems Set – Volume 18)

LEROY Maurice

Rheology, Physical and Mechanical Behavior of Materials 1: Physical Mechanisms of Deformation and Dynamic Behavior

2022

BAYLE Franck
Product Maturity 1: Theoretical Principles and Industrial Applications
(Reliability of Multiphysical Systems Set – Volume 12)
Product Maturity 2: Principles and Illustrations
(Reliability of Multiphysical Systems Set – Volume 13)

EL HAMI Abdelkhalak, DELAUX David, GRZESKOWIAK Henri
Reliability and Physics-of-Healthy in Mechatronics
(Reliability of Multiphysical Systems Set – Volume 15)

LANNOY André
Reliability of Nuclear Power Plants: Methods, Data and Applications
(Reliability of Multiphysical Systems Set – Volume 14)

LEDOUX Michel, EL HAMI Abdelkhalak
Heat Transfer 3: Convection, Fundamentals and Monophasic Flows
(Mathematical and Mechanical Engineering Set – Volume 11)
Heat Transfer 4: Convection, Two-Phase Flows and Special Problems
(Mathematical and Mechanical Engineering Set – Volume 12)

PLANCHETTE Guy
Cindynics, The Science of Danger: A Wake-up Call
(Reliability of Multiphysical Systems Set – Volume 11)

2021

CHALLAMEL Noël, KAPLUNOV Julius, TAKEWAKI Izuru
Modern Trends in Structural and Solid Mechanics 1: Static and Stability
Modern Trends in Structural and Solid Mechanics 2: Vibrations
Modern Trends in Structural and Solid Mechanics 3: Non-deterministic Mechanics

DAHOO Pierre Richard, POUGNET Philippe, EL HAMI Abdelkhalak
Applications and Metrology at Nanometer Scale 1: Smart Materials, Electromagnetic Waves and Uncertainties
(Reliability of Multiphysical Systems Set – Volume 9)
Applications and Metrology at Nanometer Scale 2: Measurement Systems, Quantum Engineering and RBDO Method
(Reliability of Multiphysical Systems Set – Volume 10)

LEDOUX Michel, EL HAMI Abdelkhalak
Heat Transfer 1: Conduction
(Mathematical and Mechanical Engineering Set – Volume 9)
Heat Transfer 2: Radiative Transfer
(Mathematical and Mechanical Engineering Set – Volume 10)

2020

SALENÇON Jean
Elastoplastic Modeling

2019

BAYLE Franck
Reliability of Maintained Systems Subjected to Wear Failure Mechanisms: Theory and Applications
(Reliability of Multiphysical Systems Set – Volume 8)

BEN KAHLA Rabeb, BARKAOUI Abdelwahed, MERZOUKI Tarek
Finite Element Method and Medical Imaging Techniques in Bone Biomechanics
(Mathematical and Mechanical Engineering Set – Volume 8)

IONESCU Ioan R., QUEYREAU Sylvain, PICU Catalin R., SALMAN Oguz Umut
Mechanics and Physics of Solids at Micro- and Nano-Scales

LE VAN Anh, BOUZIDI Rabah
Lagrangian Mechanics: An Advanced Analytical Approach

MICHELITSCH Thomas, PÉREZ RIASCOS Alejandro, COLLET Bernard, NOWAKOWSKI Andrzej, NICOLLEAU Franck
Fractional Dynamics on Networks and Lattices

SALENÇON Jean
Viscoelastic Modeling for Structural Analysis

VÉNIZÉLOS Georges, EL HAMI Abdelkhalak
Movement Equations 5: Dynamics of a Set of Solids
(Non-deformable Solid Mechanics Set – Volume 5)

2018

BOREL Michel, VÉNIZÉLOS Georges
Movement Equations 4: Equilibriums and Small Movements
(Non-deformable Solid Mechanics Set – Volume 4)

FROSSARD Etienne
Granular Geomaterials Dissipative Mechanics: Theory and Applications in Civil Engineering

RADI Bouchaib, EL HAMI Abdelkhalak
Advanced Numerical Methods with Matlab® 1: Function Approximation and System Resolution
(Mathematical and Mechanical Engineering Set – Volume 6)
Advanced Numerical Methods with Matlab® 2: Resolution of Nonlinear, Differential and Partial Differential Equations
(Mathematical and Mechanical Engineering Set – Volume 7)

SALENÇON Jean
Virtual Work Approach to Mechanical Modeling

2017

BOREL Michel, VÉNIZÉLOS Georges
Movement Equations 2: Mathematical and Methodological Supplements
(Non-deformable Solid Mechanics Set – Volume 2)
Movement Equations 3: Dynamics and Fundamental Principle
(Non-deformable Solid Mechanics Set – Volume 3)

BOUVET Christophe
Mechanics of Aeronautical Solids, Materials and Structures
Mechanics of Aeronautical Composite Materials

BRANCHERIE Delphine, FEISSEL Pierre, BOUVIER Salima, IBRAHIMBEGOVIC Adnan
From Microstructure Investigations to Multiscale Modeling: Bridging the Gap

CHEBEL-MORELLO Brigitte, NICOD Jean-Marc, VARNIER Christophe
From Prognostics and Health Systems Management to Predictive Maintenance 2: Knowledge, Traceability and Decision
(Reliability of Multiphysical Systems Set – Volume 7)

EL HAMI Abdelkhalak, RADI Bouchaib
Dynamics of Large Structures and Inverse Problems
(Mathematical and Mechanical Engineering Set – Volume 5)
Fluid-Structure Interactions and Uncertainties: Ansys and Fluent Tools
(Reliability of Multiphysical Systems Set – Volume 6)

KHARMANDA Ghias, EL HAMI Abdelkhalak
Biomechanics: Optimization, Uncertainties and Reliability
(Reliability of Multiphysical Systems Set – Volume 5)

LEDOUX Michel, EL HAMI Abdelkhalak
Compressible Flow Propulsion and Digital Approaches in Fluid Mechanics
(Mathematical and Mechanical Engineering Set – Volume 4)
Fluid Mechanics: Analytical Methods
(Mathematical and Mechanical Engineering Set – Volume 3)

MORI Yvon
Mechanical Vibrations: Applications to Equipment

2016

BOREL Michel, VÉNIZÉLOS Georges
Movement Equations 1: Location, Kinematics and Kinetics
(Non-deformable Solid Mechanics Set – Volume 1)

BOYARD Nicolas
Heat Transfer in Polymer Composite Materials

CARDON Alain, ITMI Mhamed
New Autonomous Systems
(Reliability of Multiphysical Systems Set – Volume 1)

DAHOO Pierre Richard, POUGNET Philippe, EL HAMI Abdelkhalak
Nanometer-scale Defect Detection Using Polarized Light
(Reliability of Multiphysical Systems Set – Volume 2)

DORMIEUX Luc, KONDO Djimédo
Micromechanics of Fracture and Damage
(Micromechanics Set – Volume 1)

EL HAMI Abdelkhalak, RADI Bouchaib
Stochastic Dynamics of Structures
(Mathematical and Mechanical Engineering Set – Volume 2)

GOURIVEAU Rafael, MEDJAHER Kamal, ZERHOUNI Noureddine
From Prognostics and Health Systems Management to Predictive Maintenance 1: Monitoring and Prognostics
(Reliability of Multiphysical Systems Set – Volume 4)

KHARMANDA Ghias, EL HAMI Abdelkhalak
Reliability in Biomechanics
(Reliability of Multiphysical Systems Set –Volume 3)

MOLIMARD Jérôme
Experimental Mechanics of Solids and Structures

RADI Bouchaib, EL HAMI Abdelkhalak
Material Forming Processes: Simulation, Drawing, Hydroforming and Additive Manufacturing
(Mathematical and Mechanical Engineering Set – Volume 1)

DE SAXCÉ Géry, VALLÉE Claude
Galilean Mechanics and Thermodynamics of Continua

2015

JEBAHI Mohamed, ANDRE Damien, TERREROS Inigo, IORDANOFF Ivan
Discrete Element Method to Model 3D Continuous Materials
Discrete Element Model and Simulation of Continuous Materials Behavior Set – Volume 1

JEBAHI Mohamed, DAU Frédéric, CHARLES Jean-Luc, IORDANOFF Ivan
Discrete-continuum Coupling Method to Simulate Highly Dynamic Multi-scale Problems: Simulation of Laser-induced Damage in Silica Glass
Discrete Element Model and Simulation of Continuous Materials Behavior Set – Volume 2

KARLIČIĆ Danilo, MURMU Tony, ADHIKARI Sondipon, MCCARTHY Michael
Non-local Structural Mechanics

SAB Karam, LEBÉE Arthur
Homogenization of Heterogeneous Thin and Thick Plates

2014

ATANACKOVIC M. Teodor, PILIPOVIC Stevan, STANKOVIC Bogoljub, ZORICA Dusan
Fractional Calculus with Applications in Mechanics: Vibrations and Diffusion Processes
Fractional Calculus with Applications in Mechanics: Wave Propagation, Impact and Variational Principles

CIBLAC Thierry, MOREL Jean-Claude
Sustainable Masonry: Stability and Behavior of Structures

ILANKO Sinniah, MONTERRUBIO Luis E., MOCHIDA Yusuke
The Rayleigh–Ritz Method for Structural Analysis

LALANNE Christian
Mechanical Vibration and Shock Analysis – 5-volume series – 3rd edition
Sinusoidal Vibration – Volume 1
Mechanical Shock – Volume 2
Random Vibration – Volume 3
Fatigue Damage – Volume 4
Specification Development – Volume 5

LEMAIRE Maurice
Mechanics and Uncertainty

2013

ADHIKARI Sondipon
Structural Dynamic Analysis with Generalized Damping Models: Analysis
Structural Dynamic Analysis with Generalized Damping Models: Identification

BAILLY Patrice
Materials and Structures under Shock and Impact

BASTIEN Jérôme, BERNARDIN Frédéric, LAMARQUE Claude-Henri
Non-smooth Deterministic or Stochastic Discrete Dynamical Systems: Applications to Models with Friction or Impact

EL HAMI Abdelkhalak, RADI Bouchaib
Uncertainty and Optimization in Structural Mechanics

KIRILLOV Oleg N., PELINOVSKY Dmitry E.
Nonlinear Physical Systems: Spectral Analysis, Stability and Bifurcations

LUONGO Angelo, ZULLI Daniele
Mathematical Models of Beams and Cables

SALENÇON Jean
Yield Design

2012

DAVIM J. Paulo
Mechanical Engineering Education

DUPEUX Michel, BRACCINI Muriel
Mechanics of Solid Interfaces

ELISHAKOFF Isaac *et al.*
Carbon Nanotubes and Nanosensors: Vibration, Buckling and Ballistic Impact

GRÉDIAC Michel, HILD François
Full-Field Measurements and Identification in Solid Mechanics

GROUS Ammar
Fracture Mechanics – 3-volume series
Analysis of Reliability and Quality Control – Volume 1
Applied Reliability – Volume 2
Applied Quality Control – Volume 3

RECHO Naman
Fracture Mechanics and Crack Growth

2011

KRYSINSKI Tomasz, MALBURET François
Mechanical Instability

LAMBERT Stéphane, NICOT François
Rockfall Engineering

SOUSTELLE Michel
An Introduction to Chemical Kinetics

2010

BREITKOPF Piotr, FILOMENO COELHO Rajan
Multidisciplinary Design Optimization in Computational Mechanics

DAVIM J. Paulo
Biotribology

PAULTRE Patrick
Dynamics of Structures

SOUSTELLE Michel
Handbook of Heterogenous Kinetics

2009

BERLIOZ Alain, TROMPETTE Philippe
Solid Mechanics using the Finite Element Method

LEMAIRE Maurice
Structural Reliability

2007

GIRARD Alain, ROY Nicolas
Structural Dynamics in Industry

GUINEBRETIÈRE René
X-ray Diffraction by Polycrystalline Materials

KRYSINSKI Tomasz, MALBURET François
Mechanical Vibrations

KUNDU Tribikram
Advanced Ultrasonic Methods for Material and Structure Inspection

SIH George C., NAÏT-ABDELAZIZ Moussa, VU-KHANH Toan
Particle and Continuum Aspects of Mesomechanics

Printed and bound by CPI Group (UK) Ltd, Croydon, CR0 4YY

13/09/2026

14961012-0006